NORTHSCAPES

NORTHSCAPES

History, Technology, and the Making of Northern Environments

Edited by Dolly Jørgensen and Sverker Sörlin

UBCPress · Vancouver · Toronto

21 20 19 18 17 16 15 14 13   5 4 3 2 1

Printed in Canada on FSC-certified ancient-forest-free paper (100% post-consumer recycled) that is processed chlorine- and acid-free.

Library and Archives Canada Cataloguing in Publication

Northscapes : history, technology, and the making of northern environments / edited by Dolly Jørgensen and Sverker Sörlin.

Includes bibliographical references and index.
Issued in print and electronic formats.
ISBN 978-0-7748-2571-9 (bound); ISBN 978-0-7748- 2572-6 (pbk.);
ISBN 978-0-7748-2573-3 (pdf); ISBN 978-0-7748-2574-0 (epub)

1. Arctic regions – Environmental conditions – History. 2. Arctic regions – Discovery and exploration – History. 3. Human ecology – Arctic regions – History. 4. Technology – Arctic regions – History. 5. Arctic regions – Historiography. I. Jørgensen, Dolly, writer of introduction, editor of compilation II. Sörlin, Sverker, writer of introduction, editor of compilation

G606.N67 2013   909'.09113   C2013-905265-8
C2013-905266-6

Canadä

UBC Press gratefully acknowledges the financial support for our publishing program of the Government of Canada (through the Canada Book Fund), and the British Columbia Arts Council.

UBC Press
The University of British Columbia
2029 West Mall
Vancouver, BC V6T 1Z2
www.ubcpress.ca

Contents

Illustrations

Figures

Maps

Acknowledgments

This volume grew out of the Environmental Histories of the North workshop, the third workshop of the Nordic Environmental History Network, held in October 2010. The workshop brought together thirty scholars from around the world to discuss historical approaches to the environment in this place we call "North." We thank all the participants in the workshop for their contributions, comments, critiques, and suggestions. Special thanks go to the workshop organizers, Finn Arne Jørgensen, Olof Somell, and Sverker Sörlin, for taking on the hard work of coordinating the participant travel, accommodations, and local arrangements.

We are especially grateful to the organizations that sponsored the workshop: Nordforsk, Network in Canadian History and Environment (NiCHE), and KTH Royal Institute of Technology. In addition, the Research Council of Norway sponsored Dolly Jørgensen with a two-month grant to edit this volume. KTH and Nordforsk contributed toward the production and publication costs.

Finally, we want to thank the people at UBC Press who have helped turn this scholarly endeavour into a coherent volume: Graeme Wynn, for suggesting the Press as a potential home for the book; Randy Schmidt, for championing our work through the process; and Anna Eberhard Friedlander, for getting us over all the technical production hurdles.

Dolly Jørgensen
Sverker Sörlin

NORTHSCAPES

INTRODUCTION

Making the Action Visible

Making Environments in Northern Landscapes

DOLLY JØRGENSEN and SVERKER SÖRLIN

Is there a history of the North? Since ancient times, the answer to this question has been no. History was the narrative of human action, and where human action seemed to cease in cold and ice there could be no history. This notion, derived from Hippocrates and Aristotle, who favoured the temperate zones, lived on well into the twentieth century. Alaska was a place of "cold, darkness, and monotony" wrote environmental determinist geographer Ellsworth Huntington, causing strain and inaction even on those who could have fostered "progress" in the far North. Therefore, "the man of action generally leaves Alaska because he is the one to whom the strain of inaction is least endurable during the unduly long winter."[1]

In world historiography, the North has remained of marginal importance up until recent times, largely for the same reason: the stereotype of inaction, of little being at stake. With a few exceptions, the history of northern exploration, which already a full century ago was the subject of mighty tomes and an emerging literature on economic development in the far North, served precisely as the opposite of history: a non-history of no events and the silence that preceded action.[2] The people who happened to live there, few as they were and with unclear origins, were just another example of what anthropologist Eric R. Wolf would much later call the "people without history" – perhaps one of its most perfect examples.[3]

When history, later than most other academic disciplines, discovered the North, the kind of action historians were sensitized to from their studies of action elsewhere came to light: political contestation, economic rivalry, diplomatic pursuits, violence and power, and the presence of government, Church, or military. Yet, what failed these historians was by and large a clear idea of the environment not as the restraining, deterministic factor Huntington saw it but as part of a relational human history where societies, cultures, and their natural conditions are studied in an integrated fashion.

Historians discovered the environment long before they took the North seriously, even though the idea that the environment as an integral part of history likewise entered historiography quite late.[4] The Marxist doyen of British historiography in the postwar period, E.H. Carr, wrote in his seminal Trevelyan lectures of 1961 that it was the task of the historian in "the study of man on his environment and of his environment on man ... to increase man's understanding of, and mastery over, his environment."[5] That was a brief and somewhat skewed foreboding of environmental history, which would soon burst out in earnest, starting with epic histories of environmental ideas by Clarence Glacken and Roderick Nash in the late 1960s.[6] After the Mediterranean awakening with the Annales School, a wide sweep across imperial geographies, and an array of smaller regional attempts in a number of disciplines, including geography and anthropology, as well as American geographers and ecologists writing an environmental history of sorts, the time was ripe in the 1980s and the following decades to write histories of various world regions. Environmental historians have now studied numerous regions of our planet, including Australia, the Pacific, parts of Africa, Latin America, Mexico, Southeast Asia, and even northern Europe, including Scandinavia, as well as written emerging environmental histories of entire nature types, such as oceans, mountains, marshes, forests, and deserts.[7]

Still, the North remained an empty space not just for environmental historians but for most historians. A general history of the North, or of the circumpolar regions, is yet to be found. The reason is not hard to see: apart from the old stereotype of emptiness and silence, the global North has not, until quite recently, been a region in its own right.[8] Divided into national spheres of influence and colonial possession, it has lacked unity and commonality in conventional political and economic terms. It has been perceived as not holding one history but many histories, and although these have been increasingly told in recent decades – as extensions of various strands of historical approaches, be it economic, technological, environmental, or political – virtually no one has attempted a common historical frame on a professional scholarly level.[9]

This book is an attempt to take up the challenge. Perhaps it is no coincidence that it is an attempt initiated within a network of environmental historians in a part of the world that is in itself part of the North. The idea, though raised a couple of years earlier, was adopted in the Nordic Environmental History Network (NEHN) in 2009, and a call for papers led to a selection of authors from around the northern hemisphere assembling for a three-day workshop in Stockholm in October 2010.[10] That environmental historians and historians of science and technology were keen to come together to discuss histories of the North shows that the kind of histories written in these fields have an eye to the qualities of the past that extend beyond political boundaries and are the products of human wayfaring and subtle relations between humans and the lands that surround them – qualities that can be hard to see for some but that are visible for those with the special skills to see how even the seemingly most empty space is also part of the human sphere, or the "life-world."[11] This, of course, is also a kind of action, if only you can see it. In that sense, this is a book about writing – that is, making tangible – a history that is just about to be seen; it is a work on emergence.

What Is North?

In the northern reaches of the northern hemisphere, nature poses unique challenges and provides unparalleled opportunities. Long, dark winters alternate with sun-filled summers, at least some years. Light is low, shadows long, as are the hours of dusk and dawn; in the furthest north they make up entire seasons of hazy half-light between day and night, a kind of third time. Ice, snow, and mountains seasonally restrain, seasonally facilitate inland travel; abundant seas act as highways for commerce and populations. The North is sparsely inhabited yet geographically vast, with large shares of both land and ocean. From the volcanic soils in Iceland to the boreal forests of Fennoscandia and Russia to the Canadian Shield, northern geography offers riches, but these do not come easily. Northern populations have over millennia and centuries carved out culturally and technologically functional life forms, which in turn shaped long-lasting environments. People who have come to the North in more recent times, whether they stay or move on, create distinctive northern environments of a different kind by adapting to its climatic and geographic conditions and at the same time turning landscapes to their own needs with new sets of technologies, trade, and social organization.

We recognize that North itself is not a single, definable concept. For our purposes, North is more than the Arctic (the subject of an increasing number of history volumes), but it is difficult to say where North begins on a map.[12] The attempts that have been made are not convincing, and the latitudinal limit, at any rate, varies considerably between nations.[13] In Scandinavia, with the warm Gulf Stream, North tends to overlap by and large with the official definition of the Arctic, which although defined as everything above sixty degrees north latitude, does not sit well with Scandinavian ideas of "arcticality."[14] In the continental climates of Russia and North America, North is often located more southerly than sixty degrees north latitude. Rather, North is a space imagined by people themselves, part of an identity, or state of mind, that is held not just by individuals but by institutions and entire organizations.

We have taken this approach and considered North from our historical subjects' perspectives. In spite of the difficulties of defining the geography exactly, North is a place where challenges of geography and climate are typically related to cold and inaccessibility, especially as seen from the outside, or to home, hearth, and in certain periods (like the present) even warming. Something inherent in these challenges creates similarities in the shaping of northern environments across the globe. We believe that North, then, is a valuable analytical framework that goes beyond nation-states and transgresses borders set up by previous politics and historical scholarship, which up until recent times have tended to focus on the respective North of each northern nation, mostly because each North has been ruled from a capital centre in the south.[15] This is in sharp contrast to the natural sciences, which for a long time considered, if not the entire North, at least the Arctic as a far more unified region, worthy of study for its exceptional properties and its cross-regional commonalities, an interest that has to some extent been shared by archaeologists and anthropologists but rarely by historians. That seems now to be changing, likely as an effect of the rise of the Arctic as a region of global economic and environmental significance, even going through a period of regionalization in the 1990s, a marked difference from the obscure and shrouded Cold War Arctic, which was in large parts closed off as a potential theatre of war.[16]

Our emphasis on the emergence of northern environments does not in any way overshadow the fact that the North and the Arctic have been for a long time squarely situated in geopolitical realities. On the contrary, these storylines concur; it is when the environmental significance of the North grows through colonization, resource exploitation, or climate change that strategic and political interest intensifies. Thus, in the circumpolar north, environmental history has co-evolved with economic, diplomatic, and geopolitical history to an extent that is true of few other regions in the world. This also means that the North is simultaneously a region of the world, a global asset, an indicator for climate models, a cornucopia of natural resources, and an arena for security arrangements, in particular during the Second World War and the Cold War but in our days once again with recent northern military buildup by several Arctic states.[17] These processes of global geopolitics, strategic positioning, industrial infrastructures, and physical exploitation of critical resources require technologies and science that ultimately shape the North into an environment. Modernizing technologies stand in relationship to Aboriginal technologies and knowledge systems, creating tensions that have been a dominant force, and a recurring concern, in northern societies.

Technology and the Making of Environments

We chose this volume's subtitle – "History, Technology, and the Making of Northern Environments" – to imply a creative process: technology over time actively makes environments. This contention deserves some exploration. First, we have consciously used the word "environments" rather than "nature." Although nature as a physical entity has existed in the past without human intervention and would still continue to exist should humans disappear from the universe, as long as humans are in the picture, there is no pure and untouched nature out there. As has become clear in twenty-first-century research on climate change, even the most remote parts of the globe have been and are still affected by human hands, even if the activities take place thousands of kilometres away. As a historical category of analysis, we turn instead to environments, which are produced out of the shifting relationships between humans and nature.[18] Nature can exist without people, but environments cannot; environments are places "out there" brought into the human realm. Environing is the process of domesticating nature, the moulding of places where we humans perform our actions, where we live and work and play. Thus, to understand not only what environments are but why and how they are – indeed, when they are – we need a historical perspective; we need to understand how people have historically related to northern nature to make sense of the environments we may now take for granted. Note that in this proposition we have not used the singular word "environment." This is because we realize there is not one environment – it varies over space and time. Humans have created a mosaic of environments that have mutated and shifted in response to natural forces, cultural change, economic adjustments, and technological development, to name a few of the factors involved in the process of environing.

Second, in this volume we have chosen to focus on one of the factors in the creation of environments: technology. Technology at its most basic encompasses objects, such as tools and machines, but it also includes activities – the process of design, production, and maintenance – and knowledge.[19] Technology in a sense, then, is part of all things human. It is also involved in the relations that humans have between themselves and with the natural world, regardless of whether this is described in the traditional dualistic way as humans leaving an "impact" or a "footprint" or whether you emphasize the relational and integrated relationship that the concept of environing implies.[20]

Environmental history as a discipline is well positioned to investigate the ways that northern landscapes have been read, created, modified, and even destroyed with technological tools, resulting in new and constantly changing northern environments. Interest in the intersection of technology and environmental history has been booming in the last decade.[21] Scholarship stressing technology as a primary force in creating new environments for good or bad is, however, older, as evidenced in classics of environmental history such as Donald Worster's *Dust Bowl* and Richard White's *Organic Machine*.[22] At their core, these works consider the profound changes that humans make in the non-human world through technological objects and practices, as well as knowledge. As the historian of technology Thomas P. Hughes has observed, the world is built of "intersecting and overlapping natural and human-built systems, which together constitute eco-technological systems."[23] Technology in action creates environments, and often in ways and with results that were not intended. This makes it critical for environmental historians, as well as for scholars from other environmental humanities disciplines, to critically examine the ways in which humans introducing and using technologies of all sorts intentionally or unintentionally participate in the formation of new spaces and places.

The Technology-Environment Nexus in the North

The challenges of northern nature have encouraged a historical reliance on technology to understand and overcome them. Although technologies are deployed across the globe, in northern settings these tools are often relied on to extremes. Scientific investigations of northern nature, from marine life to glaciers, have categorized, catalogued, and ordered a seemingly disordered new world of the North. Technologies – from agricultural practices to home-building techniques to travel methods – have aided people migrating to and living in extreme and very variable climates, with climate changes reinforced by the so-called Arctic amplification. People have not always been successful in their northern ventures – facing famine, environmental degradation, and their own culturally rooted maladaptation to social-ecological circumstances – but they have continued to try. Hunting and gathering nomads, migrating settlers, enduring pastoralists, Arctic explorers, and indeed even modern states have all relied on a technological toolbox to function in the North. In so doing, people have brought new northern environments into being, both in the physical sense and in the ways that they have reshaped their real and imagined life worlds and their travelled lines and itineraries through language, skills, texts, and images. Thus, the environments we refer to here are in a sense wider and deeper than the environment "out there," with all the human pursuits and their traces and relations configured into them.[24]

This volume presents how unique northern environments have resulted from the relationship between humans, technology, and northern nature. Using environmental history approaches, the chapters examine a broad range of geographies, including Iceland and other islands in the northern Atlantic, Sweden, Finland, Russia, the Pacific Northwest, and Canada, and cover a wide time span, from AD 850 to 2000 (Map I.1). The contributors are primarily historians, but chapters by an archaeologist and an anthropologist expand the methods employed to tell these stories of environing beyond the typical historical sources to archaeological digs and contemporary interviews. All of the chapters are bound together by an intellectual project of investigating North as both a shaped environment and an imagined space, particularly through the deployment of technology.

The chapters have been divided into four themes: Exploring the North, Colonizing the North, Working the North, and Imagining the North. Although the chapters could have been divided in many other ways, and they often deal with more than one theme, we think these four themes provide an overarching framework for the types of environing happening in the North.

In the first section, three chapters investigate exploration and expansion into northern environments from the eighteenth through twentieth centuries. The construction and mobilization of scientific knowledge and

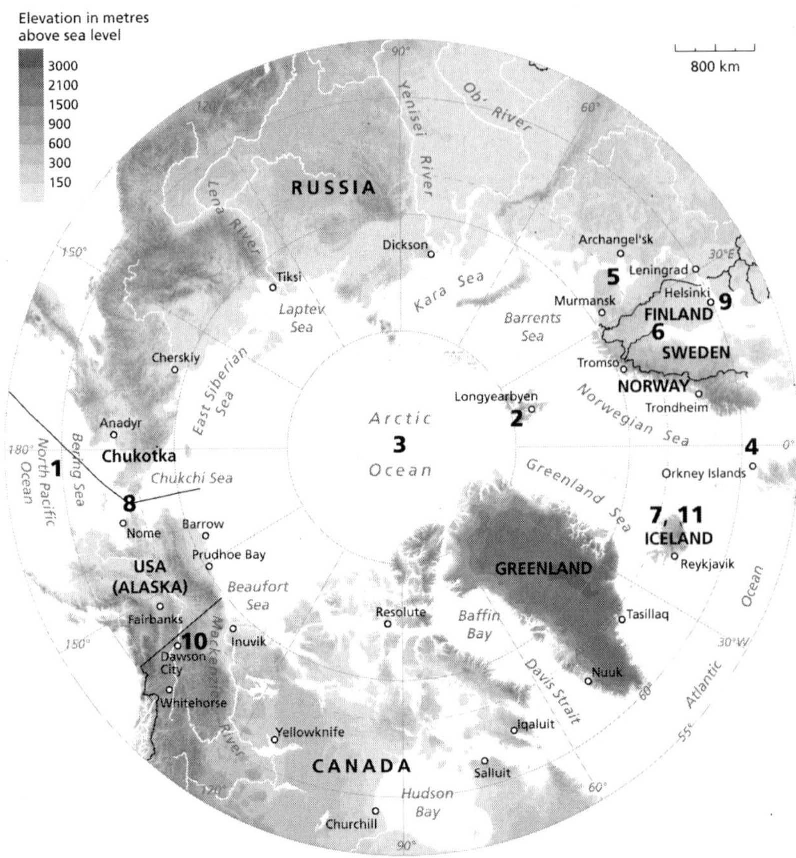

Map I.1 The geographical spread of the cases examined in the chapters in this volume, shown in circumpolar view

technological tools aided in accessing and understanding northern nature, creating opportunities for the making of new environments. As we see in the chapters by Ryan Tucker Jones and Seija Niemi, scientific exploration of the North Pacific and the fossil-rich Arctic caused scientists to rethink not only the nature of the North but also natural history at large, including climate history and evolutionary development. Northern places became critical to understanding how life has evolved, how geological and climatological processes have shaped history, and how the North fits into these global processes. These scientists employed technologies of sorting and classification to make sense of their world, which as Geoffrey Bowker and Susan Leigh Star have argued, is a ubiquitous human trait.[25]

Northern environments became a place of conquest and expansion, used both as a means of foreign policy and as a demonstration of techno-scientific prowess domestically.[26] When expanding into northern places, scientists, technocrats, and government officials alike searched for analogues and models to understand the North. In Jones's story, early biogeographers attempted to place northern species and places into worldwide patterns using technologies of categorization – stressing at first their similarity to other animals and places around the globe, then later emphasizing their differences. In Richard Byrd's adventures detailed by Marionne Cronin, we see a drive to compare the hardships of airplane travel to older transportation means in creative ways in order to maintain the idea of the heroic explorer. Comparison to the previously known, whether those references come from different spaces, times, or technologies, seems to be a common pattern in the framework of northern exploration.

As people reached out to understand northern places, they invariability reshaped those same spaces. The northern Pacific environment was irreparably modified with the extinction and near extinction of marine mammals and indigenous cultures by hunting and colonization practices that developed along with exploration. Cronin's technological approach to the history of the well-known explorer Byrd reveals how exploration *created* environments as much as *discovered* nature. The deployment of new technologies drew the North distinctly and surprisingly early into the sphere of modernity, and even served as one of its vanguard arenas.

In the second theme, Colonizing the North, two chapters focus on the ways in which people moved to the North, fashioning environments along the way. Jane Harrison, through her archaeological investigation, reveals the ways that Vikings adapted their landscapes in the North Atlantic for both practical and social ends. The Vikings brought existing building technologies with them to their new settlements but deployed them in unique ways. Julia Lajus reveals a similar pattern in Russian extension into the Murman coast region, with the Russians at first finding models for railroad and colonial expansion from Norway and Canada, then later rejecting these models in favour of their own distinctive development pattern. Technology remade the Kola North, which was criss-crossed by the railways, harvested for fish and minerals, and settled by colonists and convicts.

People in the North have exhibited great resilience and power to create livable environments, as stressed by the four chapters grouped under the theme Working the North. Social and economic landscapes were built and broken in the often harsh North, sometimes resulting in environmental disaster, as the chapter on the Icelandic sheep-raising highlands posits. The issue of Icelandic soil erosion reminds us that the environments of the North were fragile, yet historically people continued to survive and even thrive in northern spaces: Finnish and Swedish farmers modified their agricultural technological packages to cope with environmental challenges, as detailed by Jan Kunnas, and urban women faced challenges unique to northern cold climates in the story of Helsinki's washerwomen, as told by Simo Laakkonen. Technologies used to modify the living conditions of the North often had migrated from elsewhere, as we see in both the Viking mound-building technologies of the Orkney Islands and the transferred slash-and-burn and peatland cultivation agricultural techniques in the Gulf of Bothnia region. But when these technologies were employed, the environments they created were uniquely local, with social and economic consequences for local inhabitants. Bathsheba Demuth certainly reminds us of these local differences in her chapter on reindeer herding, which took very different paths in two places not spatially distant but ideologically polar opposites. It is a story of the kind of mega-schemes that states often organize, but as Demuth tells us, it is not enough to "see like a state" – in James C. Scott's sense;[27] it is also necessary for the state and, indeed, for commercial actors to write previous history out of the landscape. A placeless nature, without prior narrative, is more readily transportable and able to forge, but this comes with a cost. The technologies at work in these cases are not particularly glamorous – they are the hand tools of farmers, herders, and washerwomen – and because of that relative anonymity, they have been easy for historians to overlook as forces of environmental change. Yet, the chapters in this section show how integral they are in making northern environments.

Living in the North is also a state of mind: North is a place of *imagining* as much as it is a place of being, as the two chapters in the final section reveal. These works on Dawson City, Canada, by Lisa Cooke, and on preserving the integrity of nature in the eastern highlands of Iceland, by Unnur Birna Karlsdóttir, take us beyond the physical challenges of exploring, colonizing, and making to the ways that North can be conceptualized as home – a home of ancestors, a place of belonging, and a cultural keystone. As we see in Cooke's anthropological take on northern Canada, North is not just a place but an idea of a place that has deep roots in Canadian culture. As Cooke points out, national-cultural narratives can be "deployed as technologies in the making of this environment" as much as railroads, agricultural tools, and houses can be. Although the environment of Dawson City, with its mining history and Old West town, has been shaped in the past by physical technologies, the idea of the Klondike Gold Rush as a national heritage now has been produced through narrative technology. This turning of technological landscapes into heritage landscapes (and vice versa) is likewise the story of the battles to develop or preserve Iceland's waterfalls. On this isolated land mass in the Atlantic, North takes on special meaning because of a conjunction of Iceland's settlement history, culture, and climate. Karlsdóttir reveals how that northern natural heritage and national heritage are tightly intertwined in Icelandic ideas of hydropower development and protests against it.

Finally, Finn Arne Jørgensen brings us full circle to reflect upon what future these chapters about the past might bring. He argues that these tales reveal that North is a networked region with worldwide tendrils that extend production and consumption far beyond its own hybrid landscapes. There is a place, then, for both histories of the North and the North in history.

What these chapters collectively achieve is to demonstrate that the northern environment is not a given, fixed condition. It is a constantly changing phenomenon, moulded and shaped by societies and cultures – one that's bound to keep changing, as the region is now facing perhaps the most comprehensive resource extraction bonanza ever as fossil fuel prices rise and new sea routes open up as a consequence of climate change. Historical insights thus follow closely on the heels of current developments and could, in the best of worlds, be of use as governments, local residents, NGOs, and other actors try to navigate change in northern environments in the twenty-first century with both the advent of new technologies and the continuing productive use of many of the old ones.

NOTES

1 Ellsworth Huntington, *The Character of Races, as Influenced by Physical Environment, Natural Selection and Historical Development* (New York: Scribner's, 1924), 67.

2 One such tome is Fridtjof Nansen, *In Northern Mists: Arctic Exploration in Early Times*, Engl. trans., 2 vols. (London: William Heinemann, 1911). First published 1910.

3 Eric R. Wolf, *Europe and the People without History* (Berkeley, Los Angeles, and London: University of California Press, 1982).

4 An early seminal work was Lucien Fèbvre, *La Terre et l'évolution humaine* (Paris: Albin Michel, 1922).

5 Edward Hallett Carr, *What Is History? The George Macaulay Trevelyan Lectures Delivered at the University of Cambridge, January-March 1961* (New York: Vintage Books, 1961), 111.

6 Clarence Glacken, *Traces on the Rhodian Shore* (Berkeley: University of California Press, 1967); Roderick Nash, *Wilderness and the American Mind* (New Haven, CT: Yale University Press, 1967).

7 The sources that could be quoted to sustain this statement are too numerous to be cited here. See, however, J. Donald Hughes, *What Is Environmental History?* (Cambridge, UK: Polity Press, 2006) for a quite comprehensive coverage of major works up until then. Hughes also brings out the essential lineage of the history of the field at large. In passing, it can be observed that the Arctic plays no role in a fairly recent volume on the environmental history of northern Europe: Tamara Whited et al., *Northern Europe: An Environmental History* (Santa Barbara, CA: ABC Clio, 2005).

8 E.C.H. Keskitalo, *Negotiating the Arctic: The Construction of an International Region* (New York: Routledge, 2004).

9 There are exceptions that deal with specifically Arctic history rather than the more general North: Charles Emmerson, *The Future History of the Arctic: How Climate, Resources and Geopolitics Are Reshaping the North, and Why It Matters to the World* (London: Vintage Books, 2010); John McCannon, *A History of the Arctic: Nature, Exploration and Exploitation* (London: Reaktion Books, 2012).

10 An early instigation to the project was a conversation by Sverker Sörlin and John McNeill at the environmental and social history joint workshop in Paris in September 2008. When NEHN was formed later that same fall, the idea found an institutional home.

11 Kirsten Hastrup, "Destinies and Decisions: Taking the Life-World Seriously in Environmental History," in *Nature's End: History and the Environment*, ed. Sverker Sörlin and Paul Warde (London: Palgrave Macmillan, 2009), 331-48.

12 Recent examples of such history volumes include Graeme Wynn, *Canada and Arctic North America: An Environmental History* (Santa Barbara: ABC Clio, 2007); Roger Launius, James Rodger Fleming, and Donald H. Devorkin, eds., *Globalizing Polar Science: Reconsidering the International Polar and Geophysical Years* (New York: Palgrave, 2010); Jessica Shadian and Monica Tennberg, eds., *Legacies and Change in Polar Science: Historical, Legal and Political Aspects on the International Polar Year* (Farnham, UK: Ashgate, 2010); Sverker Sörlin, ed., *Science, Geopolitics and Culture in the Polar Region – Norden beyond Borders* (Farnham, UK: Ashgate, 2013).

13 A curious attempt to even make an index of northerliness was presented in Louis-Edmond Hamelin, *Canadian Nordicity: It's Your North Too* (Montreal: Harvest House, 1978). See also, Oran R. Young, "Bears, Boreas and Celestial Mechanics," in *To the Arctic: An Introduction to the Far Northern World* (London: Wiley, 1989), 1-24.

14 For the concept of arcticality as a northern counterpart to tropicality, see Gísli Pálsson, "Arcticality: Gender, Race, and Geography in the Writings of Vilhjalmur Stefansson," in *Narrating the Arctic: A Cultural History of Nordic Scientific Practices*, ed. Michael T. Bravo and Sverker Sörlin (Canton, MA: Science History Publications, 2002), 275-309, and David Arnold, *The Problem of Nature: Environment, Culture and European Expansion* (Oxford: Blackwell, 1995), 141-68. See also Sverrir Jakobsson, ed., *Images of the North: Histories, Identities, Ideas* (Amsterdam and New York: Rodopi, 2009), for perspectives on various meanings and interpretations of "North" in literary and artistic sources.

15 Iceland since independence from Denmark in 1918 (partial) and 1944 (full) might be a possible exception; on Iceland's modern environmental history, see Karen Oslund, *Iceland Imagined: Nature, Culture, and Storytelling in the North Atlantic* (Seattle: University of Washington Press, 2011).

16 Gail Osherenko and Oran R. Young, *The Age of the Arctic: Hot Conflicts and Cold Realities* (Cambridge: Cambridge University Press, 1989); Oran R. Young, "The Internationalization of the Circumpolar North: Charting a Course for the 21st Century," Vilhjalmur Stefansson Lecture 2000, http://www.thearctic.is (Akureyri, Iceland: Stefansson Arctic Institute); Keskitalo, *Negotiating the Arctic.*

17 A large body of work in recent years has addressed these aspects of northern history; see, for example, Ronald E. Doel, "Constituting the Postwar Earth Sciences: The Military's Influence on the Environmental Sciences in the USA after 1945," *Social Studies of Science* 33 (2003): 635-66; Keith R. Benson and Helen M. Rozwadowski, eds., *Extremes: Oceanography's Adventures at the Poles* (Sagamore Beach, MA: Science History Publications USA, 2007); Matthew Farish, *Contours of America's Cold War, 1940-1960* (Minneapolis: University of Minnesota Press, 2010); Matthew Farish, "Creating Cold War Climates: The Laboratories of American Globalism," in *Environmental Histories of the Cold War*, ed. John R. McNeill and Corinna R. Unger (Cambridge: Cambridge University Press, 2010), 51–84; P. Whitney Lackenbauer and Matthew Farish, "The Cold War on Canadian Soil: Militarizing a Northern Environment," *Environmental History* 12 (2007): 921-50; Sverker Sörlin, "Narratives and Counter Narratives of Climate Change: North Atlantic Glaciology and Meteorology, ca 1930-1955," *Journal of Historical Geography* 35 (2009): 237-55; Sverker Sörlin, "The Global Warming That Did Not Happen: Historicizing Glaciology and Climate Change," in Sörlin and Warde, *Nature's End*, 93-114; Sverker Sörlin, "The Anxieties of a Science Diplomat: Field Co-Production of Climate Knowledge and the Rise and Fall of Hans Ahlmann's 'Polar Warming,'" *Osiris* 26 (2011): 66-88; Urban Wråkberg, "The Politics of Naming: Contested Observations and the Shaping of Geographical Knowledge," in Bravo and Sörlin, *Narrating the Arctic*, 155-97; Christopher Ries, "Lauge Koch and the Mapping of North East Greenland: Tradition and Modernity in Danish Arctic Research, 1920-1940," in Bravo and Sörlin, *Narrating the Arctic*, 199-231; Matthias Heymann et al., "Exploring Greenland: Science and Technology in Cold War Settings," *Scientia Canadensis* 33 (2010): 11-42.

18 We are building here on Sverker Sörlin and Paul Warde, "Making the Environment Historical: An Introduction," in Sörlin and Warde, *Nature's End*, 1-19.

19 See the particularly insightful discussion in James C. Williams, "Understanding the Place of Humans in Nature," in *Illusory Boundary: Technology and the Environment*, ed. Martin Reuss and Stephen Cutcliffe (Charlottesville: University of Virginia Press, 2010), 9-25.

20 Concepts like impacts and footprints remain in wide currency and form the often unreflected background to overviews and textbooks such as Anthony N. Penna, *The Human Footprint: A Global Environmental History* (Malden, MA, and Oxford: Wiley-Blackwell, 2010). This is despite a mounting critique; for example, Lesley Head, "Is the Concept of Human Impacts Past Its Use-By Date?" *Holocene* 18 (2008): 373-77.

21 Many of authors writing about this intersection are featured in Reuss and Cutcliffe, *Illusory Boundary*; Susan Schrepfer and Philip Scranton, eds., *Industrializing Organisms: Introducing Evolutionary History* (New York: Routledge, 2003); and Dolly Jørgensen, Finn Arne Jørgensen, and Sara Pritchard, eds., *New Natures: Joining Environmental History with Science and Technology Studies* (Pittsburgh: University of Pittsburgh Press, 2013). Other recent examples include Sara Pritchard, *Confluence: The Nature of Technology and the Remaking of the Rhône* (Cambridge, MA: Harvard University Press, 2011); Finn Arne Jørgensen, *Making a Green Machine: The Infrastructure of Beverage Container Recycling* (New Brunswick, NJ: Rutgers University Press, 2011); Timothy LeCain, *Mass Destruction: The Men and Giant Mines That Wired America and Scarred the Planet* (New Brunswick, NJ: Rutgers University Press, 2009); Thomas Zeller, *Driving Germany: The Landscape of the German Autobahn, 1930-1970* (New York: Berghahn Books, 2007); and Thomas Zeller and Christoph Mauch, eds., *The World Beyond the Windshield: Roads and Landscapes in the United States and Europe* (Athens: Ohio University Press, 2008). SHOT has a special-interest group named Envirotech specifically for scholars investigating the intersection of technological and environmental histories. The group established an article prize, now named the Joel A. Tarr Envirotech Article Prize, in 2004. Recent winners include Christopher Jones, "A Landscape of Energy Abundance: Anthracite Coal Canals and the Roots of American Fossil Fuel Dependence, 1820-1860," *Environmental History* 15 (2010): 449-84; Robert Gardner, "Constructing a Technological Forest: Nature, Culture, and Tree-Planting in the Nebraska Sand Hills," *Environmental History* 14 (2009): 275-97; and Joe Anderson, "War on Weeds: Iowa Farmers and Growth Regulator Herbicides," *Technology & Culture* 46 (2005): 719-44.

22 Donald Worster, *Dust Bowl: The Southern Plains in the 1930s* (New York: Oxford University Press, 1979); Richard White, *The Organic Machine: The Remaking of the Columbia River* (New York: Hill and Wang, 1996).

23 Thomas P. Hughes, *Human-Built World: How to Think about Technology and Culture* (Chicago: University of Chicago Press, 2004), 156.

24 Ideas on the fine-tuned relations between body, mind, skills, and appropriations of nature in the North have been described eminently by Tim Ingold in *The Appropriation of Nature: Essays on Human Ecology and Social Relations* (Manchester: Manchester University Press, 1986) and *The Perception of the Environment: Essays on Livelihood, Dwelling and Skill* (London: Routledge, 2000).

25 Geoffrey C. Bowker and Susan Leigh Star, *Sorting Things Out: Classification and Its Consequences* (Cambridge, MA: MIT Press, 1999).

26 For the deployment of science in northern spaces as a tool of empire, see Bravo and Sörlin, *Narrating the Arctic*, and Trevor Levere, *Science and the Canadian Arctic: A Century of Exploration, 1818-1918* (Cambridge: Cambridge University Press, 1993). Non-northern areas in which science is deployed for the ends of empire have received more attention from environmental historians, as in Tom Griffiths and Libby Robin, eds., *Ecology & Empire: Environmental History of Settler Societies* (Edinburgh: Keele University Press, 1997).

27 James C. Scott, *Seeing Like a State: How Certain Schemes to Improve the Human Condition Have Failed* (New Haven, CT, and London: Yale University Press, 1999).

PART 1

EXPLORING THE NORTH

1

"A Cruel Climate without Any Kind of Art"
European Natural History and the Northern Nature of the Other Pacific, 1740-1840

RYAN TUCKER JONES

When Matvei Murav'iev took over command of the Russian colonies in Alaska in 1820, he greeted his new subjects with the message that his command would be different from that of his predecessors. In a written set of instructions distributed around the vast extent of his jurisdiction, he laid out a new pay scale and new work regulations, both for the Russians and native Alaskans. The reason for the somewhat harsher new terms, Murav'iev explained, lay with the North Pacific's difficult climate:

> Even in good climates people experience want, but here, in a poor country, in a cruel climate without any kind of art, laziness is a crime. And for that reason I rest upon this rule: whoever does not work will not be fed, and neither poverty, age, nor ailment is an excuse for laziness.[1]

Events seemed to bear out Murav'iev's pessimistic reading of North Pacific nature; the supply ship scheduled to sail from St. Petersburg failed to arrive for two years, and only an infusion of provisions from the more southerly Pacific – Hawaii and California – kept the colony fed.[2] Supplies had gotten so bad, Murav'iev reported to company directors, that he was no longer drinking tea in the morning, "although I have enjoyed this habit for 25 years."[3]

Although his intellectual debts went unacknowledged, Murav'iev's view of North Pacific nature came at the intersection of several intertwining intellectual strands that had come to define the region over the previous sixty years of European exploration and exploitation. Murav'iev's unfortunate subjects without a doubt held their own, equally important, views on northern nature, yet the most influential voices came from the European scientists and administrators who crafted new meanings for the region from 1740 to 1840. The dominant lens these men saw the North Pacific through was something then termed "natural history." Natural history, by categorizing, aggregating, and disaggregating discrete units of the natural world, served as a technology enabling Europeans to comprehend and engage northern environments. By placing the North Pacific in coherent relationships with more familiar natures, Europeans could then begin to imagine stable relationships between themselves and these environments. However, unlike modern classificatory systems, these scientists did not claim to describe only nature but freely mixed together human inhabitants (indigenous and European) as constituents of the entity they called the North.[4]

In natural historians' writings, the "northerness" of the North Pacific experienced several reconceptions; at times, writers tended to subsume the region into larger mental categories, and at other times, to cast it as a place unique in the world. Although naturalists' classifications remained "tangled and crisscrossing," some broad patterns can be discerned.[5] Initially, these men had seen the North Pacific as analogous to similar cold locations in the southern hemisphere or Greenland, and searched for correspondences between the places. In the 1770s, an increasing scientific emphasis on biogeography stressed the uniqueness of discrete locations on the earth, and the North Pacific began to look particularly cold and inhospitable. At the beginning of the nineteenth century, commentators were increasingly drawn to the supposed national character of landscapes, and the North Pacific came to be seen as more a part of Russia than a part of the North writ large.[6] In this context, Murav'iev's climate-based discipline made perfect sense. The strict oversight required in Russia's own harsh climate needed to be replicated in the similarly constituted North Pacific.

At the same time, nature too was changing as a result of European activities. Extinctions and near extinctions of animals such as the sea cow and sea otter followed the European fur trade, and deforestation began to emerge as a problem in the early nineteenth century. In this respect, the early nineteenth century's conflation of nature and nation expressed the truth that the North Pacific environment was beginning to look more like European environments. Of course, concepts of northernness did not merely reflect changing circumstances but were recruited for specific imperial ends. Russians, initially resistant to the "Russianness" of the North Pacific, ended

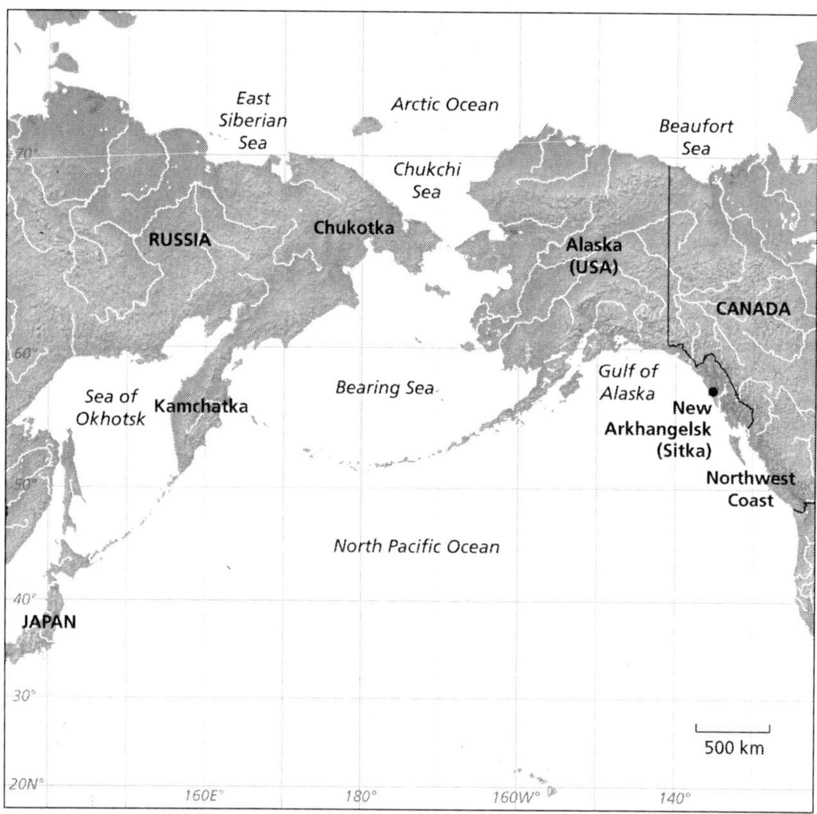

Map 1.1 The North Pacific

up using the idea to enforce labour discipline in the colonies, pointing to the grim determination needed to succeed throughout northern Russia. Thus, the idea of the North in the North Pacific helps excavate both to the concept's intellectual lineage, rooted in the European imperial experience, and to the potential contributions of the "North" to the modern imperial order.

Linnaean Natural History and Greenlandic Antecedents

To take the North Pacific as Europeans defined it around 1800 is to consider the lands, oceans, and islands from Chukotka and Kamchatka in present-day Russia to the Northwest Coast, today under the dominion of the United States and Canada (Map 1.1).[7] To engage in some twenty-first-century categorization, it is a region characterized by volcanic activity, heavy rainfall, comparatively mild winters for such a northerly region, and an abundance of marine mammals – perhaps the richest such area in the world. Today, as in the period considered here, the North Pacific is relatively thinly populated by humans and lies outside the world's major trade routes. Between 1740 and 1840, however, North Pacific nature received a great degree of attention from Europe, as new discoveries there prompted discussion and speculation about how to fit the region into prevailing notions of natural history, ethnography, and geography.

The North Pacific experienced European colonization comparatively late in the history of European overseas expansion. Not until the 1740s – when Russians first began hunting marine mammals there – were Europeans regularly visiting the region, initially without much of an interest in establishing permanent settlements. By around 1790, however, the North Pacific had become perhaps the most contested location for European imperial rivalry, one of the last attractive locations where imperial claims had not yet been solidified. Although Russia and Great Britain managed to mostly exclude other comers to the North Pacific littoral by 1840, the intervening years saw a great deal of interest in the region from all corners of Europe, from Spain to Sweden.

As the first European explorers made their way into the stormy seas and onto the rugged coasts between Sakhalin and Vancouver Island, they struggled to make sense of what they were taking in. Travellers to the North Pacific were struck first of all by the diversity and abundance of marine mammals in the sea. The ocean's pioneering naturalist, Georg Wilhelm Steller, who sailed with Vitus Bering from Siberia to Alaska in 1741-42, wrote that between Kamchatka and America "such an abundance of sea animals is to be met with that from them the costs of the most expensive expedition could be regained without much trouble in a few years."[8] Among those "Strange Beasts of the Sea" (as a translation of Steller's chapter on the animals termed them) were included the northern fur seal and a giant manatee, later named Steller's sea cow, neither of which European scientists had ever met before.

Despite the newness of the species Steller encountered, the naturalist's primary concern was to locate them within a broader schema. Steller, in fact, did not think that the North Pacific fauna differed from that in the rest of the world, but, consistent with the theories of the time, he assumed that their analogues could be found at the same latitudes in the southern hemisphere.[9] Even though he lamented that prior naturalists had let "fables and false theories after the manner of the last century and the century before" mislead them, seeing "only through a lattice what they might have seen with their own eyes," Steller himself understood the North Pacific through the lens of Linnaean biogeography.[10] In attempting to explain the geographical distribution of animals across the globe, the Swedish naturalist Carl Linnaeus had posited a uniform dispersal from a central location, probably a high island mountain that contained the altitudinal equivalents of latitudinal zones. "There therefore must have been in this island a kind of living museum," Linnaeus wrote, "furnished with plants and animals, that nothing was wanting of all the present produce of the earth."[11] As the primal floodwaters receded, animals descended from the mountains and spread out horizontally to their appropriate climatic zone. Linnaeus noted that the same zone could be found in very different locations:

> The mountains of Ararat in Armenia preserves an eternal snow upon its summit, as well as the rocks of Lapland in the Arctic circle: the same cold reigns on both; and on the tops and sides of such a mountain the same vegetables might grow, the same animals live, as in Lapland and the frigid zone; and in effect we find in the Pyrenean, Swiss, and Scotch mountains, upon Olympus, Lebanon, and Ida, the same plants which cover the Alps of Greenland and Lapland.[12]

This fairly schematic view of biogeography encouraged naturalists of the mid-eighteenth century to think in terms of worldwide correspondences for special distribution, an idea Steller readily adopted.

The assumption of animals' worldwide distribution encouraged early North Pacific naturalists to find analogues in other travellers' accounts. This quest also had the effect of occluding important environmental changes, such as extinction. A case in point was Steller's sea cow, the gigantic thirty-foot-long manatee that lived only around two uninhabited North Pacific islands (Figure 1.1). In Steller's description of the animal, published in his home town of Halle, Germany, in 1753, he was very concerned to establish the sea cow's relationship with classical and more recent descriptions of similar animals. First, Steller established that the sea cow was not the same "sea ox" that Aristotle had described, since the sea cow was far stupider than this animal. For similar reasons, the animal was likely the same that the sixteenth-century Dutch botanist and zoologist Carolus Clusius thought to be very "clumsy and ugly."[13] Many descriptions of the sea cow, however, were imperfect, some copied, and many false, according to Steller. The only real trustworthy accounts had come from the English traveller-pirate William

Figure 1.1 A Steller's sea cow, from a 1742 Russian drawing. Courtesy of Alaska State Library.

Dampier, who had encountered manatee-like creatures throughout the Caribbean, in the Philippines, and on the coasts of Australia (New Holland in his account).[14] Dampier's manatees, which lived in rivers and in shallow bays, really had very little in common with the giant kelp-eating animals that Steller observed for several months while marooned on Bering Island. Nonetheless, he reported that Dampier's animals belonged in fact to the same species observed in the North Pacific.[15] Similarly, the sea otters that Steller observed found their counterpart in the Brazilian river otters that the Dutch scientist Georg Marcgraf had recently described. Even if the descriptions did not always entirely match up, Steller's pioneering European attempt to describe the North Pacific's natural world emphasized the region's correlation with other places in the world.

Those who became interested in the North Pacific sea cow after Steller's death found themselves enmeshed in a bog of murky speculation about the animal's uniqueness. The British zoologist Thomas Pennant accepted the link Steller had made with Dampier's manatee and then extended the confusion. Although admitting that the genus was found nowhere else in the northern hemisphere, he was convinced that the "white-tailed manati" was the same species in accounts from the South Indian Ocean island of Rodrigues collected by the British naturalist Joseph Banks. Pennant also suspected that the sea cow inhabited waters all the way from the Indian Ocean to the island of Mindanao, in the Philippines, though he had no accounts of sea cows between the China Sea and the Aleutian Islands. Why and how the animal's wide range should be interrupted with a four-thousand-mile gap went unexplained.[16] The sea cow's supposed discontinuous distribution details just how little naturalists of the eighteenth century thought of a region containing a specific flora or fauna and were instead inclined to see local species as manifestations of a global type, even in the face of logical inconsistency.

The assumption of the northern sea cow's global presence lessened the impact of reports of the animal's extinction. Russian hunters, following in the wake of reports of the abundant marine mammals to be found in the North Pacific, decimated the sea cow population, which in any case had been fairly small. It is likely that the animal was extinct by 1768. However, because of widespread philosophical opposition to the possibility of extinction, and, just as importantly, to the assumption that other sea cows were still to be found in other parts of the world, few naturalists took notice of the increasing number of reports that the animal was in fact gone. As late as 1835, naturalists were still writing that sea cows could still be found, perhaps off California, or perhaps further north in Alaska.[17]

A particularly important reference point for Europeans interested in the North Pacific was Greenland. Steller again initiated the comparison. He had read Adam Olearius's account of a visit to Schleswig (modern Germany), where the author had encountered several Greenlandic Eskimos versed in whaling. Steller is thought to have applied what he had learned in the account to devise a method for capturing sea cows on Bering Island.[18] Others at St. Petersburg's Imperial Academy of Sciences, such as Gerhard Friedrich Müller, who read and interpreted Steller, noted also that the same sorts of whales and other marine mammals found around Greenland had been seen in the waters between Kamchatka and Alaska.[19] William Coxe, an English reverend who came to St. Petersburg in the 1770s, picked up Steller's thread and wove it into a picture of the North Pacific as a mere extension of Greenland. Coxe perused the Russians fur traders' accounts of travel to the Aleutian Islands and decided that they had been missing a crucial geographical link:

> [The Alutiiq] call themselves Kanagit, a name that has no small resemblance to Karalit; by which appellation, the Greenlanders and Esquimaux on the coast of Labradore distinguish themselves; the difference between these two denominations is occasioned perhaps by a change of pronunciation, or by a mistake of the Russian sailors, who may have given it this variation.[20]

Coxe cited other apparent linguistic similarities between the disparate regions, building what he considered an airtight case for cultural homogeneity across the breadth of the American North. Such evidence, besides renewing hope that a Northwest Passage connecting Greenland and Alaska might still be found, suggested that the two regions shared some essential qualities.

The Greenland that Steller, Müller, and Coxe had in mind came largely from the writings of the Norwegian missionary Hans Egede. In 1729, Egede wrote his immensely influential *Det gamle Grønlands nye perlustration* (later translated as *Description of Greenland*) with the intent of convincing the Danish state to further the island's colonization. Egede spent much of his time in Greenland searching for the lost Norse colony, a quest that resonated with those interested in the lost seventeenth-century Russian colony rumoured to have survived somewhere in Alaska. Such concerns reflected a more general, worldwide European search for "lost" people (e.g., Prester John in India, Chinese mandarins in the Caribbean, the Lost Tribes of Israel everywhere from Virginia to New Zealand). Certain people, as much as animals, were expected to have a worldwide distribution.

Egede's observation of Greenland's nature had consequences for European conceptions of the North Pacific. Despite painting Greenland in the most favourable light possible, Egede had to concede that "compared with other Countries, it is but very mean and poor." However, this poverty provided all the more reason for Denmark to lift the colony up through improvement. Although admitting that agriculture showed small prospect of success, Egede saw the ocean as Greenland's bright hope. "But as to the Seas," he wrote,

> they yield more Plenty and Wealth of all sorts of Animals and Fishes, than in most other Parts of the World, which may turn to very great Profit; witness the exceeding Riches many Nations have gathered, and are still gathering from the Whale Fishery, and the Capture of Seals and Morses or Sea Horses.

Although his ideas for a Greenland colony would not find immediate success, Egede had in fact written the blueprint for the profitable colonization of Arctic lands. Where the ideal tool of colonization – agriculture – had failed, the capture of marine mammals would have to suffice. "Thus it is confessed," he concluded, "*Greenland* is a Country not unworthy of keeping and improving."[21]

The establishment of a successful colony based on the hunting of marine mammals had no precedent then. Europeans had hunted whales in the North Atlantic for nearly three centuries, but besides temporary bases on Newfoundland and Spitsbergen, no permanent colonies had grown up around these industries. Moreover, the hunting of seals and other marine mammals such as Egede referred to had never been prosecuted on a large scale. The French killed seals around their Quebec colony but only as an ancillary activity. Of course, codfish had been a sufficient marine resource on which to found New England colonies, but this was a hunt with far different parameters, well known from European precedent. In this context, Egede's work was eye-opening for scientists and imperial promoters, and it illustrates how natural history could work as technology for unlocking hidden relationships between humans and the environment.

The idea that Arctic locations could finance profitable colonies had clear implications for the small numbers of Russians then sailing through the North Pacific in search of fur seals and sea otters. In these poor, desperate bands of men Russian imperial administrators soon perceived the beginning of empire. The notion of Alaska's resemblance to Greenland gained a strong hold on later colonial officials and the fur traders themselves. Like Coxe, the Siberian merchant and founder of the Russian American Company, G.I. Shelikhov, perceived a similarity between the Alutiiq and Greenlanders.[22] Natural historians of the 1820s considered the Alutiiq, Aleut, and Greenlandic languages to be of the same family.[23] In the 1830s, governor of Russian America Ferdinand von Wrangel, noted that

> the Russian colonies have little in common with other European settlements in America; one may possibly compare them with the Danish possessions on the west coast of Greenland. The climatic conditions – regions constantly shrouded in fog, either timbered or swampy, or completely barren, rocky and mountainous – are utterly unsuited to arable farming or stockbreeding ... Therefore [the inhabitants] are completely dependent upon the sea for sustenance.[24]

Wrangel's comments indicate that, by his time, the similarities between Greenland and the North Pacific had obtained a greater degree of environmental specificity, though the notion that the sea provided the regions' only dependable riches remained from Egede and Müller's conceptions. That, in fact, was the prevailing notion of North Pacific by the last quarter of the eighteenth century: a region similar to other places around the world with like climates that contained little of interest to the European except the wealth of the sea.

James Cook and New Ideas of the "North" Pacific

The Russian near monopoly on North Pacific exploration was broken by the British in the 1770s, when the latter renewed the search for the Northwest Passage and entered the sea otter trade themselves. Captain James Cook's third voyage, in particular, brought the North Pacific to the attention of the rest of Europe. Other Europeans (and Americans) followed, either with Cook-like voyages of exploration, or with commercial ventures. Cook's voyages, which so dramatically influenced European conceptions of the Pacific, also had an impact on environmental categories.

The German naturalist Johann Reinhold Forster, who sailed with Cook on the second voyage, took advantage of his globe-spanning experiences to come to some "philosophical" conclusions about the patterns apparent in the natural world. Forster significantly refined Linnaeus's conception of species distribution, noting that the species diversity seemed to decline with increasing latitude. Whereas in the tropics the naturalist observed a "luxuriance of vegetation," at the poles, the harsh frosts "almost precludes the germination of plants."[25] At the same time, Forster thought that northern polar climates were milder than the southern hemisphere's polar regions, mainly because the former contained far more land than did ocean-wreathed Antarctica. Thus, while Forster painted the world in broad strokes, he also began to fill in the detail that would render the North Pacific a unique location in the world.

Others took up Forster's biogeographical conception, adding several wrinkles to the idea of a barren North. As Thomas Pennant (the same Welshman who introduced so much generality into sea cow taxonomy) wrote in his study of northern animals, *Arctic Zoology*, in 1792,

> It is worthy of human curiosity to trace the gradual increase of the animal world, from the scanty pittance given to the rocks of *Spitzbergen*, to the swarms of beings which enliven the vegetating plains of *Senegali*: to point out the causes of the local niggardness of certain places, and the prodigious plenty in others. The Botanist should attend the fancied voyage I am about to take, to explain the scanty herbage of the *Arctic* regions; or, should I at any time hereafter descend into the lower latitudes, to investigate the luxuriancy of plants in the warmer climes.[26]

Figure 1.2 The sea otter had become a symbol of the North Pacific by the nineteenth century. Drawing by John Webber, 1785. Courtesy of Alaska State Library.

In Pennant's view, the "niggardness" of the Arctic climes was compensated by the abundance of useful animals in the North. His idea of animal distribution, which assumed a literal Mount Ararat instead of Linnaeus's imprecise mountain island as the point of origin, relied on divine wisdom to guide animals to the places where they would be most useful to humans.[27] Kamchatka, for example, had received an abundance of fish as Providence's recompense for the peninsula's lack of grain and cattle.[28] This was an idea picked up by Russian commentators, who also echoed the pious invocations of the Creator's hand in explaining the North Pacific's bounteous marine life, referencing especially the usefulness of these animals' furs for humans trying to survive in a cold climate.[29]

Pennant's biogeography therefore allowed for more idiosyncrasy and concerned itself less with balance than its predecessors. Pennant wondered how animals had migrated to different places – why, for example, there were reindeer found in Greenland but none native to Iceland, as the two had similar climates.[30] This train of thought led him to place particular emphasis on the range of distribution of each animal he discussed. The sea otter, for example was "the most local" animal found in the *Arctic Zoology*, meaning that it inhabited the North Pacific and nowhere else (Figure 1.2). In the wake of

Cook's decisive dismissal of the southern continent and George Vancouver's slaying of the Northwest Passage, Pennant banished Steller's associative biogeography to the discredited, overly theoretical musings of the past, much as Steller had tried to banish the fables of his predecessors.

Pennant's (and, contemporaneously, E.A.W. Zimmerman's similar *Geographische Geschichte des Menschen und der vierfüßigen Thiere*)[31] pairing of species with delimited geographical spaces encouraged Europeans to think of the North Pacific as a place with a unique nature. It was, as Pennant had written, the only place where the sea otter was to be found. Russians in the nineteenth century would speak of Alaska as the "realm of the sea otter," and it was *de rigueur* for visitors to the colonies to come back with some scrap of sea otter fur as a present or memento.[32] Pennant's work helped travellers understand the implications of overhunting in the North Pacific as well. Martin Sauer and Georg Heinrich von Langsdorff, both of whom sailed on Russian voyages to the North Pacific, brought the *Arctic Zoology* with them. When confronted with reports that the sea cow had disappeared from those waters, the men used Pennant's biogeographical theories (if not his erroneous accounts of Philippine sea cows) to claim that the animal was in fact entirely extinct. In addition, they issued the first warning that the sea otter could meet a similar fate, for reasons of its limited distribution.[33]

In 1803, the Danish geographer Conrad Malte-Brun put a postscript to the era of theoretical biogeography. "It is not yet certainly demonstrated," he wrote, "that any species of animals, perfectly identical, has been distributed by nature over all the regions of the globe. In similar climates the organizations have assumed characters which nearly approximate, but never exactly coincide."[34] His biogeography took into account a broad range of factors, including varying climates across the same latitudes, the geographical obstacles to dispersal, and even the historical effects of hunting on animal populations. Beavers, for example, "once perhaps native of all the countries of the globe," had by this time been reduced by human persecution to a far narrower range.[35] The North had become many Norths, which furthermore were subject to revolutions.

Besides reshaping biogeographical conceptions, Cook's (and others') voyages of the late eighteenth century inaugurated a general European fascination with the Pacific, which Alan Frost has called the "eighteenth century's new world." As heavily romanticized descriptions of the South Pacific came back to Europe, the polar regions – often visited by crew only recently enjoying all the delights of Tahiti – began to seem more gloomy and forbidding by contrast.[36] The barren, snow-bedecked mountains of the polar seas

Figure 1.3 A view of Snug Corner, Prince William Sound, Alaska, by John Webber, 1785. Courtesy of Alaska State Library.

especially seemed to irritate the European eye, despite a contemporaneous enthusiasm for Alpine landscapes. Gazing on an Antarctic island, Forster was revolted by the "continual fogs" and mountains "absolutely covered with ice and snow ... and in all probability incapable of producing a single plant." Nature at the ends of the earth was "too tremendous for mortal eyes to behold. The mind indeed, still shudders at the idea, and eagerly turns from so disgusting an object."[37] Typical of European impressions of the region, John Webber's depiction of Prince William Sound stresses the barrenness and inhuman scale of North Pacific landscapes (see Figure 1.3). Sailing directly from Hawaii and Tahiti to the Northwest Coast in 1778, Cook and his men also found the North Pacific hugely disappointing. "A more dreary prospect I never yet came in the way of," Charles Clerke wrote upon seeing Kamchatka (a place much modern literature praises for its incredible beauty) for the first time.[38]

Adelbert von Chamisso, a German poet-naturalist sailing with a Russian voyage in the 1820s, made the contrast between North and South Pacific natures more explicit. After many months in the South Pacific, the Russian voyages headed to Chukotka and Kamchatka, and Chamisso's mood turned philosophical. Having "cast a look over the waters of the Great Ocean [the Pacific], and its shores, and viewed the islands situated in it, between the tropics," it was now time to consider the North Pacific:

> We now turn from those gardens of pleasure to the dreary north, in the same ocean basin. The song is past. A clouded sky receives us on the very limits of the northern monsoon. We penetrate through the gloomy veil, which eternally hovers over these seas, and shores not shaded by a tree, inhospitably frown upon us, with their snow-crowned summit. We shudder to find man also settled here![39]

Lest Chamisso's disgust be taken for a general condemnation of the North, later he noted that European nature at similar latitudes was far more benevolent than "the melancholy portrait of these coasts." Norway and Lapland, in particular, came off well in Chamisso's comparison, presenting scenes comparable to Italian gardens or "one of the lakes of Switzerland."[40] The Spanish naturalist Jose Mariano Mozino, on the other hand, thought the climate of the Northwest Coast of America far more benign than at other North American regions at a similar latitude.[41] Although there was little consensus in such matters of taste, by the 1820s, European aesthetic preferences had begun to mirror biogeographical conceptions by demarcating distinct northern natures, some pleasing and some horrific.

National Nature

While biogeographers located species specific to the North Pacific, others were busy determining the biogeography of another animal – humans. In the late eighteenth and early nineteenth centuries, commentators increasingly made connections between North Pacific nature and the national "characters" of the region's colonizing powers, usually based on a very vague notion of climatic affinity. Perhaps the most bombastic example could be found in German playwright August von Kotzebue's 1798 play about Count Maurice Benyowsky. Benyowsky, a Hungarian (or Polish – the story was even then clouded in mystery) prisoner of war exiled to Siberia, had executed a daring and bloody escape from Kamchatka some twenty-five years previous, gaining widespread fame in the process. Benyowsky's story held great promise for contemporary playwrights interested in probing its moral implications, or just interested in a sanguinary tale in an exotic location. Benyowsky's story, it turned out, was also good for thinking about North Pacific nature.

Von Kotzebue's *Benyowsky*, which was performed from St. Petersburg to Washington, DC, put a romantic spin on the revolutionary's anarchic spirit. Von Kotzebue infused the play's action with philosophical musings in which "nature" played a large role in shaping the protagonists' actions. Much of the story's message turned on a supposed connection between the natural conditions of the North Pacific and the Russian Empire. Athanasia, daughter of the Kamchatka garrison commander and Benyowsky's love interest, contrasts the benevolence of European climes with the dreariness of Kamchatka: "Italy is a lovely country ... There orange groves blossom; here we work them in tapestry. There nature is a healthy youth; here an infirm grey-headed old man."[42] Ironically, Kamchatka and the North Pacific as a whole were, in geological terms, in fact some of the youngest parts of the world, constantly being reshaped by volcanic forces. Still, nature as old and infirm corresponded to prevailing views of Russia's supposed resistance to change and its love of ancient institutions such as autocracy.

Benyowsky himself emerged in von Kotzebue's play as the most eloquent mouthpiece for a growing association of Russian nature with Russian despotism. Upon arrival in Kamchatka the desperate exile proclaims, "Human might has combined with all-powerful nature to thwart us. On this side, desert wastes and boundless fields of snow – on that, trackless seas bar us from the habitable world." Through Benyowsky's companion, the old exile Crustiew, the playwright also managed to recruit South Pacific images, imbuing them with the liberty that Russian nature lacked. Crustiew looks out to the sea and declares wistfully, "To fly! to fly to the Isles of Marian! ... The Island Tinian – A terrestrial paradise! Free! free! a mild climate! a new created sun! harmless inhabitants, wholesome fruits – and liberty!"[43] As another romantically inclined author, the Frenchman Barthélemy de Lesseps wrote, Benyowsky "escaped from the mountains of snow, under which the Russians supposed him to be buried."[44] Whereas Europeans had commonly associated Siberia with political exile, now nature itself was taking on the characteristics of the Russian state.[45] As the region's nature shed its cosmological associations with other northern places, it began to acquire a new personality that was linked as much to politics as anything else.

As the picture of North Pacific nature grew gloomier, predictions of Russian success in the region grew rosier. Vancouver, who visited the North Pacific in 1792, thought the Russians the only people capable of establishing an empire there, because of their own background in a "frigid region."[46] In addition, the German naturalist Alexander von Humboldt considered the North Pacific inimical to Spanish climatic proclivities but thought the region would be a comfort for Russian colonists:

> The progress of the Siberian Russians toward the south naturally will be more rapid than that which the Mexican Spanish will make toward the

> north. A hunting people, accustomed to living under cloud-filled skies in an excessively cold climate finds agreeable the temperature reigning on the coast of New Cornwall [Pacific Northwest]. And this same coast, on the other hand, seems an uninhabitable land, a polar region, to colonists coming from a moderate climate, from the fertile and delicious plains of Sonora and New California.[47]

Russian imperialism in the North Pacific took on such a natural hue that the geographer Malte-Brun proposed calling Alaska "American Siberia," hardly a compliment to either place, since he described them as regions "where nature sees her vivifying influence expire, and witnesses the awful termination of her vast empire."[48] Malte-Brun's implicit juxtaposition could serve as a fitting encapsulation of the North Pacific's transformation – where nature's empire receded, the frigid, despotic Russian Empire took its place.

Although Russian Enlightenment-era writers had been hostile to Western attempts to link Russian climate and culture,[49] in the first decades of the nineteenth century, Russians began to accept an association with northern nature. Those at home celebrated Russia's "northernness" in verse and visual arts.[50] Peter Simon Pallas, Russia's most gifted scientific mind of period, produced a massive zoology that saw the North Pacific's remarkable biodiversity as a microcosm of the Russian Empire's own ethnic and geographical diversity.[51] This changing orientation was also manifested in a nascent revolt against the prevailing European fascination with the South Pacific. In the eighteenth century, Russians had been less influenced by comparisons with the South Pacific than were other Europeans, not having taken part in the early exploration of the tropics. After 1802, when their ships first circumnavigated the globe and visited the South Pacific, Russian voyagers still did not find the South Seas as enchanting as had the British and French. The fiercely patriotic Fedor Shemelin, who sailed with the first Russian circumnavigators, remarked with surprise and disgust after visiting the Marquesas:

> besides a lot of mice, [South Pacific Islanders] don't have any animals whose skins would be worth even a kopek ... and, while they may eat pigs on holidays, the rest of the time they feed themselves on seeds, molluscs, and fish. And the fish are not even of the tasty kind; just bonitos, mackerel, etc.[52]

In the North Pacific, on the other hand, there was an "amazing abundance" of salmon, which according to Shemelin, were far tastier than the Pacific's tropical fish.[53] Russian naturalists observed that tropical waters contained very little animal life, but northern waters were rich in molluscs and crustaceans, as well as marine mammals.[54] Sometimes this preference for northern nature struck a resigned, rather than proud, note. In 1818, officials wrote back to the Russian American Company's (which administered the Alaskan colonies for the empire) Board of Directors in St. Petersburg, discussing the failure of recent imperial ventures in Hawaii and California: "God grant us that the north will open its treasure to us. The south is not so benevolent."[55] Northern nature had finally found a dedicated patron in Russia.

The growing distinction between North and South Pacific natures shaped Russian policy in its colonies. The environment's supposed effect on human behaviour could lead to a number of conclusions, often to the detriment of northern peoples. While in the Aleutians, Russian explorer Otto von Kotzebue was surprised to find the people happy and mirthful, acting more like "lively South Sea Islanders, instead of the serious inhabitants of the north."[56] Most of Russia's conquered Pacific subjects were seen as cheerless but hardy workers. A comment in the Russian American Company archives bemoaning the difficulty of recruiting labourers to the North Pacific colonies is telling. Whereas Russians and Aleuts handled the cold just fine, "it is laughable to think of California or Hawaiian Islanders working here, for the Company has no power over them. Having been born in a warm climate, they cannot tolerate the cold."[57] Such remarks buttressed the movement toward Murav'iev's desultory exhortations (found at the start of this chapter) driving his subjects to hew a hard living out of the frigid wastes. In 1839, the eminent St. Petersburg naturalist K.E. von Baer outlined one of the stranger implications of life in the harsh North Pacific climate. Regarding supposed Russian atrocities toward the area's indigenous inhabitants, "it is difficult to distinguish how great is the guilt which was borne by men and how many of the reported evils were the unavoidable result of the injurious climate."[58] North Pacific nature thus served as justification both for Russian colonization – based on supposed climatic affinities – and for the grimness of the struggle to wrest riches out of a hostile environment, a task that often demanded brutalities people from softer climes did not have to employ.[59] Here natural history worked as a technology for constructing a labour regime more suited to the climatic realities it proposed to reveal.

Of course, as Russians and others imputed a national character to the North Pacific, the region's distinctiveness, as crafted by the biogeographers, receded like the outgoing tide. The supposed Russian qualities of the North Pacific, as "American Siberia," overwhelmed its distinctive flora and fauna.

This intellectual movement in fact reflected the region's ongoing environmental history. European imperial expansion did tend to homogenize local ecosystems. Overhunting of fur-bearing animals, halting attempts to plant and harvest European grains, and some localized deforestation all meant that the North Pacific had begun to resemble Europe in small but increasing ways. As a London periodical remarked in 1815, soon those exotic seas would be just another European North Sea, "where a seal is so great a rarity as to be called a mermaid."[60] Steller's sea cow, as we have seen, had already descended into the marine mammal underworld, to reappear again, like a fattened and bewhiskered mermaid, only in sailors' legends.

The history of northern nature in the North Pacific from 1740 to 1840 can be seen as a movement in three acts. First, Europeans struggled to find analogues with which to help them understand the region, a process that emphasized the North Pacific's similarity to other northern natures. Second, new theories of biogeography encouraged a closer look at North Pacific nature, and a sharper, more distinctive picture emerged in accounts from the late eighteenth century. Third, a new nineteenth-century interest in linking nature and nation caused Europeans to impute to those islands and seas the character of the Russian Empire, a trend that reattached the North Pacific to other locales. Imperial interests can be seen in each of these conceptions – mid-eighteenth-century comparisons with Greenland encouraged Russia to envision a colony based on the exploitation of marine mammals, and the late-century pessimism about North Pacific nature helped Europeans dismiss the validity of Aboriginal cultures. However, the conflation of nature and nation reflected most clearly the imperial order that had solidified by 1840. The Russian nature of the North Pacific littoral naturalized the extension of Russian labour practices.

As discussed in other chapters in this book, in later eras, humans turned increasingly to narratives of heroism, work, and technology to define the North. Eighteenth- and early-nineteenth-century Europeans, however, showed little concern with these categories, demonstrating instead a greater need to classify the vast Pacific spaces in terms of natural history. This broad discipline did not hesitate to incorporate humans, animals, and climate into its classificatory systems, preferring to conflate rather than separate them. In retrospect we can recognize the process of categorization itself as a technology used to reveal appropriate human relationships with northern environments. As the North Pacific moved through different categories, those categories suggested different human actions. This same process continues in the twenty-first century as the North Pacific loses its (fairly recent) national associations and merges back into a general "North," thought to be uniformly subject to particularly violent climate change and uniquely vulnerable to human activity in general.

NOTES

1 M.I. Murav'iev, "Prikaz," 4 November 1820, in *Rossiisko-Amerikanskaya Kompaniya i izuchenia tikhookeanskogo severa, 1815-1841*, ed. N.N. Bolkhovitinov (Moscow: Nauka, 2005), 96.

2 P.A. Tikhmenev, *A History of the Russian American Company*, trans. Richard Pierce and Alton Donnelly (Seattle and London: University of Washington Press, 1978), 213.

3 M.I. Murav'iev, "Donosenie," 12 November 1823, in Bolkhovitinov, *Rossiisko-Amerikanskaya Kompaniya*, 159.

4 On classificatory systems, see Bowker and Star, *Sorting Things Out*, 46. Some eighteenth- and nineteenth-century natural historians disagreed altogether about the need to classify discrete parts of nature, but they were part of a small and declining faction within the discipline. See Harriet Ritvo, *The Platypus and the Mermaid: And Other Figments of the Classifying Imagination* (Cambridge, MA: Harvard University Press, 1998), chap. 1.

5 Quotation from Bowker and Star, *Sorting Things Out*, 21.

6 The "national character of landscapes" classificatory technique bears a resemblance to what Michael Bravo terms "anthropogeography," though mid-nineteenth-century natural historians in the North Pacific postulated a relationship between humans and nature owing far more to supposed political characteristics than environmental ones. See Michael Bravo, "Measuring Danes and Eskimos," in Bravo and Sörlin, *Narrating the Arctic*, 240.

7 Quotations from G.W. Steller, *Journal of a Voyage with Bering*, ed. O.W. Frost, trans. Margritt Engel and O.W. Frost (Stanford, CA: Stanford University Press, 1988), 66-67; Peter Simon Pallas, *Betrachtungen über die Beschaffenheit der Gebürge ...* (Frankfurt und Leipzig, 1778), 42; Otto von Kotzebue, *A Voyage of Discovery, into the South Sea and Beering's Straits*, vol. 3 (London: Longman, Hurst, Rees, Orme, and Brown, 1821), 283; Ferdinand von Wrangel, *Russian America: Statistical and Ethnographical Information*, trans. Mary Sadouski (of 1839 German edition) (Kingston, ON: Limestone Press, 1980), 86.

8 G.W. Steller, "Letter to Gmelin, 1745," in *Bering's Voyages: An Account of the Efforts of the Russians to Determine the Relation of Asia and America*, vol. 2, ed. Frank Golder (New York: American Geographical Society, 1922-25), 245.

9 Georg Wilhelm Steller, *De Bestiis Mariniis, or the Beasts of the Sea*, trans. Walter Miller (Lincoln: University of Nebraska Press, 2005), 61.

10 Ibid., 39.

11 Quoted in Janet Browne, *The Secular Ark: Studies in the History of Biogeography* (New Haven, CT, and London: Yale University Press, 1983), 18.

12 Carolus Linnaeus, *Dissertation II: On the Increase of the Habitable Earth*, in *Foundations of Biogeography: Papers with Commentaries*, ed. Mark V. Lomolino, Dov F. Sax, and James H. Brown (Chicago: University of Chicago Press, 2004), pt. 5-8, 14.

13 G.W. Steller, *Ausfuehrliche Beschreibung von Sonderbaren Meerthieren* (Halle, Germany, 1753), 51, 93.

14 William Dampier, *A New Voyage Round the World*, vol. 1 (London: James Knapton, 1707), 64.

15 Steller, *Ausfuehrliche Beschreibung*, 102.

16 Thomas Pennant, *Arctic Zoology*, vol. 1 (London: Henry Hughs, 1784), 179.

17 Tilesius von Tilenau, "Die Wallfische," *Isis von* Oke, vol. 28 (Jena, 1835), 719, 720.

18 O.W. Frost, "Adam Olearius, the Greenland Eskimos, and the First Slaughter of Bering Island Sea Cows, 1742: An Elucidation of a Statement in Steller's Journal," in *Russia in North America: Proceedings of the 2nd International Conference on Russian America, Sitka, Alaska August 19-22, 1987*, ed. Richard Pierce (Kingston, ON: Limestone Press, 1990), 123-28; David Scheffel, "Adam Olearius's 'About the Greenlanders,'" *Polar Record* 23, 147 (September 1987): 701-11.

19 G.F. Müller, "Von Dem Walfischfang in Kamtshcatka," in *Die Grosse Nordische Expedition*, ed. Wieland Hintzsche and Thomas Nickol (Gotha, Germany: Justus Perthes Verlag, 1996), 246. Such ideas were also current in British natural historical circles. Thomas Falconer wrote to Joseph Banks in 1768 that there was a "chain of similitude" from Germany to Kamchatka. Quoted in Bernard Smith, "European Vision and the South Pacific," *Journal of the Warburg and Courtauld Institutes* 13, 1/2 (1950), 67.

20 William Coxe, *Russian Discoveries between Asia and America* (London: J. Nichols, 1780), 116.

21 Hans Egede, *A History of Greenland: Shewing the Natural History, Situation, Boundaries, and Face of the Country* (London: Printed for C. Hitch, 1745), 4, emphasis in original.

22 Kirill T. Khlebnikov, *Notes on Russian America*, trans. Marina Ramsay (Kingston, ON, and Fairbanks: Limestone Press, 1994), pt. 2-4, 4.

23 See Ivan Veniaminov, *Notes on the Islands of the Unalashka District*, trans. Lydia Black and R.H. Geoghegan (Kingston, ON: Limestone Press, 1984), 293-94.

24 Quoted in Alix O'Grady, *From the Baltic to the Pacific* (Kingston, ON, and Fairbanks: Limestone Press, 2001), 188-89.

25 Johann Reinhold Forster, *Observations Made on a Voyage Around the World* (London: G. Robinson, 1778), 169. Forster also gained significant insights into island biogeography.

26 Pennant, *Arctic Zoology*, vol. 2 (London: Hughes, 1785), B1.

27 Ibid., B2, 254.

28 Pennant, *Arctic Zoology*, vol. 1, 222.

29 Khlebnikov, *Notes on Russian America*, 3.

30 Pennant, *Arctic Zoology*, vol. 1, 25.

31 E.A.W. Zimmerman, *Geographische Geschichte des Menschen und der vierfüßigen Thiere* (Leipzig, 1778-83). Zimmerman's work was probably more influential than Pennant's, and Pennant took some of his inspiration from the German's work. However, Zimmerman had little knowledge of or concern for the North Pacific.

32 Richard Pierce, ed., *The Russian American Company: Correspondence of the Governors* (Kingston, ON: Limestone Press, 1984), 139-40.

33 See Ryan Jones, "'A Havock Made among Them': Animals, Empire, and Extinction in the Russian North Pacific, 1741-1810," *Environmental History* 16 (2011): 585-609.
34 Conrad Malte-Brun, *Universal Geography: Or, a Description of All Parts of the World on a New Plan*, vol. 1 (Boston: Wells and Lilly, 1824), 520.
35 Malte-Brun, *Universal Geography*, 521.
36 Alan Frost, "The Pacific Ocean: The Eighteenth Century's 'New World,'" in *Studies on Voltaire and the Eighteenth Century* 152 (1976): 779–822.
37 Forster, *Observations Made on a Voyage*, 168.
38 James Beaglehole, *The Journals of Captain Cook*, vol. 3 (London: Hakluyt Society, 1967), 642.
39 Adelbert von Chamisso, "Remarks and Opinions of the Naturalist of the Expedition, Adelbert von Chamisso," in Otto von Kotzebue, *A Voyage of Discovery*, vol. 3, 261.
40 Ibid., 271, 302. Chamisso quotes from Leopold von Buch, *Reise durch Norwegen und Lappland* (Berlin: Nauck, 1810).
41 Jose Mariano Mozino, *Noticias de Nutka: An Account of Nootka Sound in 1792*, trans. Iris Higbie Wilson (Seattle: University of Washington Press, 1970), 5.
42 August von Kotzebue, *Count Benyowsky; or, the Conspiracy of Kamtschatka*, trans. from German by Rev. W. Render (London: J. Deighton and J. Nicholson, 1798), 3.
43 Ibid., 36, 37.
44 Barthélemy de Lesseps, *Travels in Kamtschatka, during the Years 1787 and 1788*, vol. 1 (London: J. Johnson, 1790), 157.
45 Galya Diment and Yuri Slezkine, *Between Heaven and Hell: The Myth of Siberia in Russian Culture* (New York: St. Martin's Press, 1993).
46 George Vancouver, *A Voyage of Discovery to the North Pacific Ocean*, vol. 5 (London: John Stockdale, 1805), 343.
47 Alexander von Humboldt, *Political Essay on the Kingdom of New Spain*, vol. 2 (London: Longman, Hurst, Rees, Orme, and Brown, and H. Coburn, 1811), 387.
48 Malte-Brun, *Universal Geography*, 154.
49 See, for example, Catherine II's response to the Frenchman Abbe Jean Chappe d'Auteroche's condemnation of Siberian nature and Russian despotism; Anonymous [Catherine II], *The Antidote: Or an Enquiry into the Merits of a Book, Entitled a Journey into Siberia* (London: S. Leacroft, 1772).
50 See Christopher Ely, *This Meager Nature: Landscape and National Identity in Imperial Russia* (DeKalb: Northern Illinois University Press, 2002).
51 Peter Simon Pallas, *Zoographia Rosso-Asiatica* (St. Petersburg: Caes. Academiae Scientiarum Impress, 1831).
52 Fedor Shemelin, *Zhurnal Pervogo Puteshestviia Rossiian vokrug Zemnogo Shara ...*, vol. 2 (St. Petersburg: Meditsinskaia Tipografiia, 1816-18), 235.
53 Ibid., 236. See also Ryan Jones, "Lisiansky's Mountain: Changing Views of Russian America," *Alaska History* 25 (2010): 1-22.
54 Frederic Litke, *A Voyage Around the World, 1826-1829*, vol. 1, *To Russian America and Siberia* (Kingston, ON: Limestone Press, 1987), 41.
55 Richard Pierce, ed., *The Russian American Company*, 20.
56 Otto von Kotzebue, *A Voyage of Discovery, into the South Sea and Beering's Straits*, vol. 1 (London: Longman, Hurst, Rees, Orme, and Brown, 1821), 211.

57 Correspondence of 17 June 1821 (mo. 449), in Bolkhovitinov, *Rossiisko-Amerikanskaya Kompaniiai*, 255.
58 K.E. von Baer, "Introduction," in Wrangel, *Russian America*, xvii.
59 Although the Virginian colonist William Byrd claimed in the late seventeenth century that the mildness of some climates meant that imperial masters had to work conquered subjects even harder. "Surely there is no place in the World," he wrote, "where the Inhabitants live with less Labour than in N Carolina. It approaches the nearer the Description of Lubberland than any other, by the great felicity of the Climate, the easiness of raising Provisions, and the Slothfulness of the People." See Paige Raibnon, "Naturalizing Power: Land and Sexual Violence along William Byrd's Dividing Line," in *Seeing Nature through Gender*, ed. Virginia Scharff (Lawrence: University Press of Kansas, 2003), 24-25.
60 *London Monthly Review*, May-August 1815, 129.

How Fossils Gave the First Hints of Climate Change
The Explorer A.E. Nordenskiöld's Passion for Fossils and Northern Environmental History

SEIJA A. NIEMI

Fossils are fundamental pieces of the geological history of the earth. Most of our knowledge about the life of previous eras of geological time is derived from fossils. According to American paleontologist Steven M. Stanley,

> The organisms that have inhabited the earth in the course of geological time have left a partial record in the rock of their presence and their activities. This record reveals that life has changed dramatically during the history of the earth and that its transformation has been intimately associated with changes in physical conditions on earth – in climates or in the position of continents, for example.[1]

Numerous books, articles, and reports have been written about fossils from the point of view of the natural sciences such as ecology, biogeography, biology, and palaeobotany; so far, I have not found any discussion of fossils by environmental historians and only a small number of studies connecting them to environmental history in the geosciences.[2]

Fossils were the special interest of Adolf Erik Nordenskiöld (1832-1901), a Finnish-Swedish explorer and scientist, mineralogist, geologist, and paleontologist. He made ten expeditions (1858-83) to the Arctic in the initial stages of the fossil discoveries. He described his explorations in several publications in several languages, significantly enlarging the existing scientific knowledge of the desolate polar regions.[3] Nordenskiöld sent many of his numerous fossil findings to Swiss professor Oswald Heer, in Zurich, who was one of the pioneers in palaeobotany. As a turning point in scientific understanding of global history, their fruitful cooperation created some of the first discussions on significant climatic changes that had occurred dozens of millions of years before.[4]

Nordenskiöld's steady opinion was that fossils play a fundamental role in explaining early climate changes. From fossils and the techniques of nineteenth-century natural science he discovered information about Earth's origins and the natural phenomena of the ancient environments, as well as about extinctions of animals and vegetation in the past. The finding of fossil mammals in northern Europe – elephants, rhinoceroses, and others now limited to tropical climates – suggested drastic climatic changes.

Although Nordenskiöld has a fundamental role in northern environmental history, there are rather few scientific studies on his expeditions in the Arctic regions and none on his fossil findings.[5] He is best known within the history of cartography owing to, for instance, his charting of Spitsbergen and the north coast of Siberia and his remarkable collection of historical maps that he gathered later in his life.[6] Nordenskiöld's writings on his Arctic expeditions are fruitful sources for learning how early scientists constructed their knowledge of the Arctic and of the world, particularly global climate change history.[7]

A Northern Scientist of the Nineteenth Century

Adolf Erik Nordenskiöld was born in Finland in 1832. His great-great-grandfather, Johan Nordberg, had moved from Sweden to Finland at the end of the seventeenth century. The Nordenskiölds – including Adolf's father, Nils Gustaf Nordenskiöld (1792-1866), the chief director of mining in Finland (1823-55) – were interested in literary activities and natural scientific issues.[8] Nils took the young Adolf on mineralogical excursions and taught him to identify and collect minerals. Thus, the boy developed a sharp and steady eye for identifying minerals, expertise he found quite profitable later on in his career.[9]

Nordenskiöld studied chemistry, natural history, mathematics, physics, mineralogy, and geology in the Imperial Alexander University in Finland. He earned his bachelor's degree in the spring of 1853. In the autumn he accompanied his father to the Urals on a mineralogical survey in the iron and copper mines of Prince Demidov in Tagil [Nizhne-Tagilsk]. He made numerous mineral findings that later on were the elemental factors of his scientific analyses. In spring 1854, he returned home, went on with his

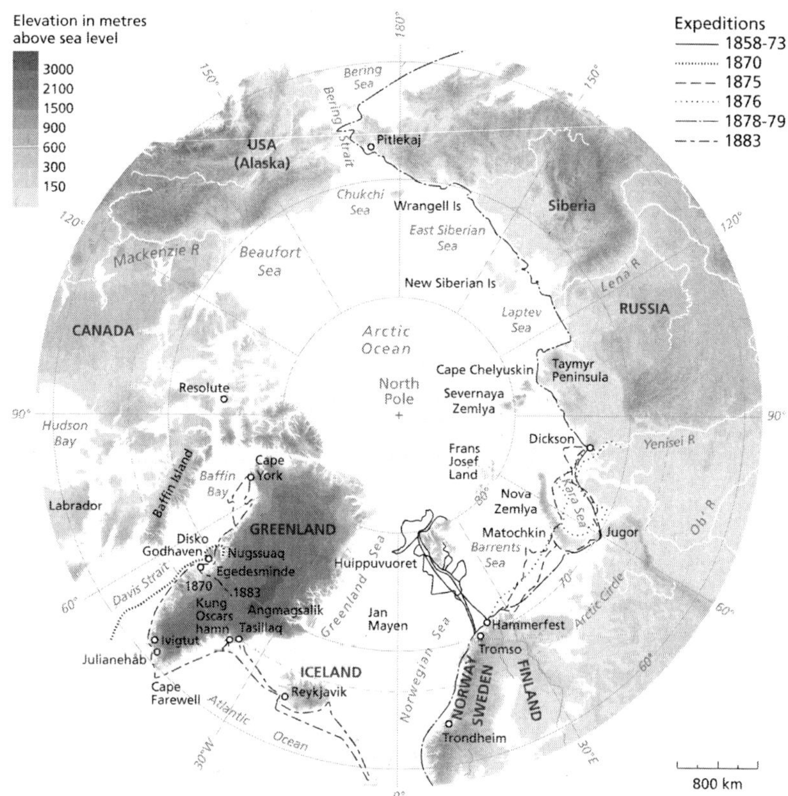

Map 2.1 Adolf Erik Nordenskiöld's expeditions

studies, and published his licentiate dissertation *Grafitens och Chondrotitens kristallformer* (Crystal Formations of the Graphite and Chondrotites).[10]

During the spring and early summer of 1856, Nordenskiöld conducted mineral analytical studies in Berlin, in the laboratory of his father's good friend, Professor H. Rose.[11] As long as his father lived, they carried on frequent correspondence and discussed, among other things, mineralogical subjects and fossils. For instance, from Berlin, Nordenskiöld wrote about his visit to Rüdersdorf's extensive chalk mine, where he had found a large number of mussels, and about another visit to Hennigsdorf, with its petrifactions from the Triassic period. He preferred to identify the rock formations by the content of fossils rather than by chemical analysis.[12]

In 1857, Nordenskiöld was forced to leave his native country of Finland because of his liberal-minded political opinions. He moved to Stockholm,

Sweden, and quite soon was appointed as director of the Department of Mineralogy at the Swedish Museum of Natural History in Stockholm and professor in mineralogy at the Royal Swedish Academy of Sciences. His Arctic explorations were tightly bound with the Academy of Sciences and the Swedish Museum of Natural History, and he made his scientific expeditions under the Swedish flag (Map 2.1). He never returned to live in Finland. He died in 1901 at his summer residence, Dalbyö, Sweden. He considered himself a Swedish national with a Finnish heart.[13]

What Did Fossils Reveal to Nordenskiöld?

Fossils were a quite recent subject within natural science in Nordenskiöld's time. The technologies of categorizing and delineating among different historical epochs were in a formative stage. Between 1822 and 1850, the main sequencing systems, including the Cambrian, Cretaceous, Triassic, Silurian, and Devonian, had been scientifically described and generally accepted by the scientific community (see Figure 2.1). The first formally recognized

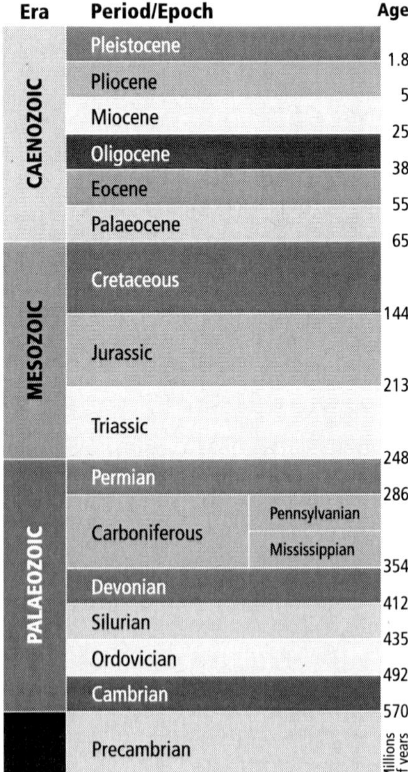

Figure 2.1 The currently accepted geological time scale

Palaeogenic epoch, the Eocene, was established in 1833 by Scottish geologist Charles Lyell on the basis of deposits found in the Paris and London basins.[14] Lyell named and described the Eocene Series in his book *Principles of Geology* (1830-33), a work that served to popularize the uniformitarian view of geology. Lyell's great discovery was that all observed geological changes could be explained by the slow, cumulative effects of the same natural forces that we can observe today, rather than by the earlier theory of catastrophic events such as the deluge.[15]

In the first half of the nineteenth century, geological and paleontological activity become increasingly well organized with the growth of geologic societies and museums and an increasing number of professional geologists and fossil specialists. The word "paleontology" was invented in 1822 by French naturalist and zoologist Georges Cuvier (1769-1832), who also is known for founding the science of paleontology. Cuvier strongly supported the fluvial theory and opposed ferociously any notion of evolution, attributing the extinction of species and the discovery of fossils of animals that have disappeared to major environmental catastrophes.[16]

The first petrifactions in the Arctic were found only a few decades prior to Nordenskiöld's work. For instance, in Spitsbergen, the English rear admiral Sir William Edward Parry had discovered the first mountain limestone petrifactions in 1827, and Sven Lovén discovered Jurassic fossils in 1837. Nevertheless, according to Nordenskiöld, most of the earliest specimens were collected under unfavourable conditions and by non-specialists. Fortunately, he and his colleagues were professional scientists who knew how to improve the scattered knowledge. Nordenskiöld, for instance, discovered Tertiary (subperiod of the Caenozoic era) plant remains in 1858; Christian Wilhelm Blomstrand found the Tertiary strata in 1861; and Otto Martin Torell, Anders Johan Malmgren, and Blomstrand identified the post-Tertiary beds, containing *Mytilus* mussels, in 1861. One valuable discovery for the geological history of Spitsbergen was saurian reptile strata and Devonian fish remains found by Nordenskiöld in 1864.[17]

The novelty of Arctic fossil finds led Nordenskiöld to write on the subject with great enthusiasm throughout his whole explorer career. For example, in his very first letter, on 29 August 1858, after returning from his first Arctic expedition to Spitsbergen, he excitedly wrote about the fossils to his mother:

> The scientific results achieved both by Torell and me are quite considerable. I found limestone beds (that means the lowest bed of the Carboniferous period) containing plenty of petrifactions. Even in later formations of limestone

> or brown coal which did not contain so many petrifactions could be found wood, plant replicas, leaf impressions, remains of fruits etc. which had been transformed into silicon and, evidenced the richness of the vegetation up here. All this for Daddy's account.[18]

The last sentence expresses the gratitude he felt for his father, who had taught him to discover fossils.

In his first fossil findings, Nordenskiöld found evidence of a totally different climate compared with the Arctic's contemporary icy environment. In a very long letter to his father, he describes how calm weather and headwind forced the ship to drop anchor in the north harbour at Bell Sound. In the end, this compulsory stop turned out to be a lucky event, since Nordenskiöld found nearby loose stones containing beautiful, big petrifactions of brachiopods from genera *Produktus*, *Spirifer*, and *Terebratula*, which he presumed to be from the lowest formation of the Carboniferous period. The rocks contained beds of sandstone, chalk, silicon, and flints. The sandstone bed had no fossils, but the other beds included rich and quite well preserved remains of ancient animals, including *Euomphalus*, tails of *Cidaris*, and even some examples of brachiopods. He found petrified molluscs of the species *micula*, *leda*, *astarte*, and *unio*, which he presumed to be from a quite recent geological period. In the Ice Fjord, at Green Harbour, he found a thin bed containing petrified tree trunks, root fibres, fruits, gravel, and pieces of snail shell, among other things. The pieces of trees were relatively well preserved, and he was able to identify the barks and the cambium. The fossils allowed him to better understand the systems of nature and its functions in previous geological periods.[19]

Nordenskiöld hoped to get an answer to his big questions: "What kind of transitional stages existed between the Miocene period and contemporary glacier formations?" and "What kind of animals and plants could have flourished around the pole during the Miocene and the European Ice Ages?" However, since he never found any unbroken series of fossil remains from every geological era, he presumed that the evidence had been crushed and pushed away by the glaciers. But on his fifth expedition to Spitsbergen in 1872-73, Nordenskiöld returned to the same areas in Green Harbour that he had visited in 1858, and this time he found a colossal fossil herbarium on Cape Lyell where every leaf told of ancient amazing vegetation corresponding to the contemporary southern flora.[20]

From the Spitsbergen fossils Nordenskiöld was able to deduce that the present climate partition was formed during the most recent geological period and that the prevailing contrast between the equatorial and polar climates was not possible during the period just prior to our time. For instance, some references indicated that the beds of the *Mytilus* mussels on Spitsbergen were much younger than the Ice Age beds in Scandinavia. That meant that Spitsbergen was not frozen during the European Ice Age. The earlier warmer climate around the pole and the cold of the Ice Age were not consequences either of a higher middle temperature or some sudden universal cold period, but rather provided a different climate partition compared with the present. It also indicated that a steady and almost equal climate had several times alternated with sharply divided temperature zones. Since he found neither smoothed stones, erratic blocks, nor any other beds resembling the contemporary Scandinavian Ice Age beds, Nordenskiöld concluded that Spitsbergen had not often been covered by ice during earlier geological periods.[21]

Fossilized plants gave Nordenskiöld insights into the transformation of vegetative regimes over time. On his two expeditions to Greenland in 1870 and 1883, Nordenskiöld made careful observations on different species, and with the help of his natural science techniques he could detect alterations that might have been passed over by a less careful researcher. He found over twenty sources of petrified plants belonging not only to the thoroughly examined Miocene formations but even to the Cretaceous period, as well as samples from two other geological periods, one between the Cretaceous and Miocene periods, the other later than the previously known Miocene strata. Out of the twenty-seven species he collected from Puilasok, on the south coast of Disko Island, a third were equal with the lower Miocene species in Greenland, a third were types not found in Greenland but which existed in the Miocene strata in Europe, and a third were new species, indicating transformations within the Miocene vegetation. Nordenskiöld was eager to inform his colleagues of these important new findings and, in a letter to his wife, asked her to pass the news to his friends at the academy.[22]

Animal remains also interested Nordenskiöld. He made expeditions to the north coast of Russia, including twice to the Yenisei River, and the first voyage ever through the Northeast Passage in 1878-79. The rich harvest of Siberian animal and vegetation remains from the Carboniferous period corresponding with the recent findings in Scandinavia inspired him greatly, as it gave him a lively picture of the subtropical vegetation of the ancient world. The remains of the mammoths with flesh and hair conserved occupied his thoughts, since questions such as "Did the mammoths live in the same period as the Scandinavian Ice Age?" and "What kind of flora and fauna existed then?" were still unsolved. His opinion was, however, that Siberia during the European glacial period had about the same climate as at present, and that the glaciers in Europe depended only on local circumstances.[23]

A careful examination of the petrifactions showed that an extensive polar continent had risen and sunk in the ocean several times, and periodical alterations of warm and cold climates had occurred. In November 1875, Nordenskiöld published an article in English, titled "On the Former Climate of the Polar Regions," in *Geological Magazine* arguing that the geology of the polar tracts supplies us with information in two different ways: "partly by comparison of the fossil animals and plants found with existing forms that live under certain determinate climatic conditions, partly by an accurate examination of the various strata of different geological ages." In another article from the next year, "Sketch of the Geology of the Ice Sound and Bell Sound, Spitzbergen," he published the results of the Swedish polar expeditions. He emphasized that Spitsbergen is one of the most interesting spots on the globe from a geological point of view. He considered that probably during the glacial period Spitsbergen's west coast was the west coast of a considerable Arctic continent, which was connected with Scandinavia in the south and with continental Siberia in the east. He also noted that the glaciers in Recherche Bay in Bell Sound had considerably advanced since the French expedition in 1838; a harbour where whalers anchored a couple of centuries back was completely filled up by a glacier, and similar advancing glaciers occurred in a great many other places. Of course, he also discussed fossils. In the Brachiopoda fossils, which are disturbed in nearly all the countries where mountain limestone is found, he noticed a close connection between the mountain limestone of Spitsbergen and Russia. The final result of his examinations was that the strata from which the Brachiopoda fossils were derived belonged to a division of the mountain limestone of the Carboniferous formation but possessed a character different from the intermixture of species occurring only in the Permian formation in other countries. This led to his supposition that "the Mountain Limestone of Spitsbergen was a later link, if not corresponding, at least analogous to the Upper Mountain Limestone of Scotland, which is separated from the Lower Mountain Limestone by a series of Coal-beds."[24] This link would have remained unnoticed by someone who was not as careful an observer as Nordenskiöld.

Nordenskiöld compared the various sediments to the pages in the geological history of the polar countries to reveal important and useful knowledge about the history of the earth.[25] Gradually, Nordenskiöld learned to see at a glance the best places for fossil findings. In the Arctic regions, the scant vegetation leaves the landscape barren, making it easier to strike and research the beds there than those in other regions. However, the structure of the ground is mostly loose sandstones or sand, neither of which is an ideal rock type for fossils, leading many non-professional fossil hunters to dig deep holes without finding anything.[26] With Nordenskiöld's approach, the North was an abundant fossil fountain; on almost every step one could find information from ancient times.

Networked Science and Fossils

None of Nordenskiöld's expeditions or science took place in a vacuum. He was an integrated part of a network of scientists who discussed each other's data and even examined each other's fossils. Sometimes their investigations corroborated theories, while at other times, contradictory ideas came to light.

In Sweden, Nordenskiöld's first promoter and guardian was Sven Lovén, "a reasonable man," as Nordenskiöld described him.[27] Lovén was one of the pioneers of invertebrate animal research. He was a member of the Royal Swedish Academy of Sciences and worked as a director at the Department of Invertebrate Zoology at the Swedish Museum of Natural History. He too was a keen fossil collector and, in 1837, on his exploration of Spitsbergen, he made an important discovery when he found a link between the contemporary Arctic fauna and that of the Scandinavian Ice Age. In 1858, he recommended Nordenskiöld to Swedish geologist and zoologist Otto Martin Torell (1828-1900) for an expedition to Spitsbergen. Torell was one of the first Swedish scientists to embrace the glacial theory, and together, Torell and Nordenskiöld would continue the research Lovén had started.[28]

The glacial theory was one of the hottest scientific subjects at the beginning of the nineteenth century. For the majority of scientists, the biblical flood had been the cause of the current configuration of the globe's surface and the formation of sedimentary rocks. For them, it was obvious that the flood had transported large boulders of rock, called erratic blocks, and composed numerous hillocks of rock debris, called moraines. Scottish geologist Charles Lyell formed a transport theory according to which transportation of minerals happened either directly by mechanical effect of water or indirectly by the transport of the rock blocks on "ice rafts." Other scientists realized these blocks could also be pushed by the ice, as can be observed in present-day glaciers. The Swiss glaciologist, paleontologist, and geologist Louis Agassiz (1807-73) had in the beginning real doubts but, gradually, the field facts led him to affirm that, during the Ice Age, a gigantic ice sheet covered large parts of the northern hemisphere as far as the Mediterranean. He succeeded in convincing other scientists, among them Charles Lyell, who soon became one of the best advocates of the glacial theory.[29]

Another theory that had found many advocates was proposed by Mr. James Croll, who argued that "the winters during the Glacial Period happened in aphelion, when ... the Earth was eight and half millions of miles further distant from the Sun during winter than it is at present."[30] The British geologist F.W. Hutton tried to validate Croll's theory in a 1875 *Geological Magazine* article by stating that this theory had found great favour because it was supposed to rest on astronomical evidence. But he had also found that, in reality, the astronomical evidence was, if anything, slightly against it: "The theory is founded on speculations in Meteorology, a science not even so well understood as Geology. The theory of the change in the obliquity of the ecliptic, advocated by Lieut.-Colonel Drayson, Mr. Belt, and others, is simply a supposition which is altogether opposed by astronomy." The editors of *Geological Magazine* referred, in the footnote, to Nordenskiöld: "A more serious difficulty for Geologists has arisen than that of explaining the cold of the Glacial Epoch; namely, to explain the warm temperate and even subtropical heat of the Earlier Tertiary periods in high northern latitudes. Such changes are not 'founded on conjecture.' See Professor Nordenskiöld's article in the November Number of the Geol. Mag. p. 525."[31] Nordenskiöld thus became integrated into the glacial debate through his work on fossils.

In 1877, a reader of *Geological Magazine,* A.H. Green, asked about Professor Nordenskiöld's arguments on recurrent glacial periods, since he had noticed that Professor Judd had written in the magazine "repeatedly of late, not without some flourish of trumpets, how completely Professor Nordenskiöld has demolished Mr. Croll and his theory of the causes of glacial epochs." Again, the editors of *Geological Magazine* referred to the articles in question[32] and stated that Nordenskiöld's papers had "good material in [them] bearing on the former climate and the extinct floras."[33]

Nordenskiöld was even more directly involved in the question of kryokonite, which he discovered along with Sven Berggeen during their trip to the Greenland ice cap in 1870. Berggeen, a botanist, hardly expected to find any flora on the ice, but surprisingly he observed, partly on the ice, partly among fine gravel, a brown multicellular alga in circular cavities, from one to three feet in depth, not only near the shore but also far inland. Nordenskiöld examined the appearance of the substance in its relation to geology and demonstrated that "it cannot have been washed down from the mountain ridges at the sides of glaciers, as it was found evenly distributed at a far higher elevation than that of the ridges on the border of the glaciers, as well as in equal quantity on the top of the ice-knolls as on their sides or in the hollows between them." It had been neither distributed over the surface of the ice by running water nor pressed up from the hypothetical bottom "ground" moraine. Therefore, the clay must "be a sediment from the air, the chief constituent of which is probably terrestrial dust spread by the wind over the surface of the ice," and "cosmic elements exist in this substance, as it contains molecules of metallic iron which could be drawn out by the magnet, and which under the blowpipe gave reaction of cobalt and nickel."[34] Nordenskiöld named this dust "kryokonite," from the Greek *kruos* (ice) and *konis* (dust). He also deducted that although the alga is very tiny, it is, with the gravel and various other microscopic organisms, the worst enemy of the enormous ice cap. The dark mass of alga absorbs much more of the sun's warming rays than the white ice and drills deep holes all over the ice, accelerating its melting. He also concluded that the same vegetation once created the same kind of ice melting in northern Europe and North America during the Ice Age.[35]

Nordenskiöld's arguments garnered great scientific interest. The cosmic elements particularly aroused different opinions.[36] For instance, the Swedish geologist Nils Olof Holst published a report on his geological investigations in which he also discussed kryokonite.[37] He had noted that it quite often "formed in little balls as big as beans, which readily absorb heat from the sun, causing the under laying ice to melt."[38] He compared his analysis of the dust with that of Nordenskiöld's and received similar results. However, he refused to believe any cosmic origin for the dust and referred to other scientists – A. von Lasaule, F. Zirkel, and E. Swedmark – who had made their own microscopical analyses on the kryokonite and all agreed "in the main: *The kryokonite contains nothing but the ordinary components of primitive rock.*"[39] Kryokonite (now usually called cryoconite) has received growing interest in the twenty-first century.[40] The common conclusion is that although some cryoconite is of cosmic origin, "it can be defined as any fine-grained material transported by wind onto ice surface."[41] Most contemporary researchers concentrate on the biological part of the dust, but some also discuss cryoconite's mechanics, concluding in general that cryoconite must absorb more energy than the glacier surface itself in order to grow.[42]

Scientists often did not have the necessary technological equipment or knowledge to analyze all of their samples themselves, encouraging the formation of scholarly networks. Nordenskiöld most often sent his fossil samples to the paleobotanist Oswald Heer. Professor Heer argued, by a careful examination of the rich materials accessible, and by a comparison of the petrifactions with those of the same period found in more southerly locations, that considerable variety of climate existed already on the face of the earth in the Miocene era, and the pole enjoyed a climate fully comparable with that of central Europe at present. Because the vegetation of Europe had an almost entirely American character, he supposed that the two continents were once united, bounded on the south by an ocean extending from the Atlantic over the present deserts of the Sahara and central Asia to the Pacific.[43] In his book *Flora Fossilis Arctica*, Heer advanced his idea that all the floral types of the more southern latitudes originally grew on "a great continuous Miocene continent within the Arctic Circle" and that from this centre they spread southward in a radial manner.[44] Some of the trees and other plants of the Miocene and Cretaceous periods in Greenland were also growing in Europe. Certain species exist even today, but in more southern areas of the earth. Of the 133 species found in Greenland, 50 are also familiar in Europe. Of the Miocene trees and bushes, 22 species are common in Greenland and Europe, and even in North America.[45] Heer's opinion was that, with such an abundance of observations and scientific material such as those Nordenskiöld brought home, our knowledge widened significantly. However, Nordenskiöld was typically not satisfied with his career because his long-term goal had been to reach the highest possible northern latitude or even the North Pole, which he failed to achieve despite his many attempts. Heer cheered him up by stating that his achievements within the field of paleontology were much more important than access to the highest north.[46]

Although Nordenskiöld usually sent his fossil findings to Heer, he sometimes sent examples to other scientists. For instance, on his third expedition, in 1864, he passed on skeletons of a large crocodile-like animal of the Tertiary period, which he had found at the Ice Fjord, to English biologist Thomas Henry Huxley. Charles Lyell said in a letter to Nordenskiöld that Huxley considered them "very interesting & seem to belong to two species of *Ichthyosaurus* one of which is new. This is very important in so northern a locality."[47]

The Royal Society stated that the Arctic expeditions that had produced the most satisfactory results in determining the nature of the fauna and flora of the Arctic Cretaceous and Miocene strata were those undertaken by E. Whymper and R. Brown in 1867, Nordenskiöld in 1870, and J.J.S. Steenstrup in 1874, when Disko Island and Noursoak Peninsula were carefully examined. There also was a reference to numerous plant fossils of which a copious list by Oswald Heer was furnished, and a mention to Nordenskiöld's opinion that "the eruptions which gave rise to these vast beds of basalt took place after the commencement of the Cretaceous and ceased the end of the Tertiary period."[48] Nordenskiöld's research was integrated and fundamental to nineteenth-century theories about Arctic climate change through the millennia.

Nordenskiöld's influence among the contemporary scientific circles upon the polar issues was quite relevant. He improved the scattered information of the earlier explorers and, in turn, younger scientists after him counterpointed his views and analysis. He was a vivid link in the chain of the cumulative knowledge of the Arctic fossils and former climates. He had a deep passion for plant fossils, which he considered to be better leads to former climates than marine animal fossils.[49] The question of cryoconite is certainly one of the most interesting issues Nordenskiöld left for the following generations of scientists. And some of the knowledge of the Arctic fossils he gathered is still relevant today.

Nordenskiöld could be rather stubborn if there was something he really wanted. For instance, since the Swedish Museum of Natural History in Stockholm had no special department for the botanical fossils Nordenskiöld brought home, they were kept in the mineralogical department. It took ten years, from 1874 to 1884, for Nordenskiöld to convince the Academy of Science and the Swedish government that a separate department was needed. His good colleague Alfred Nathorst was the first director of the palaeobotanical department from 1884 onward. Nordenskiöld was considered to be one of the authorities of the polar expeditions. His contributions in the international media were acknowledged, and he also published his scientific results in popular Swedish newspapers, thus widening the knowledge of the Arctic among the public.[50] Historian Ralph O'Connor has observed that although today "some find it difficult to see what was so appealing about nineteenth-century geological writings, for all that they were bestsellers in their day." The texts of Nordenskiöld were read by an increasingly varied public, even though they might have seemed like "islands of evocative prose separated by oceans of fossil descriptions, lengthy discussion of strata, lists of minerals, and osteological analyses." To nineteenth-century readers, the moments of epiphany were windows into a brave new world. Through them we can glimpse the indescribable sensations underlying the whole, "as the geologist – even the most articulate of geologists – gropes for the appropriate language."[51]

Perceptions of Northern Nature

The Arctic was for Nordenskiöld an empty space far from the inhabited and civilized world. With the help of the geology of the inaccessible polar regions he could observe the geological phenomena similar to the other parts of the world, such as the movement of the glaciers, the impact of cold and warmth, and the uplift of the ground. The minimal vegetation made it easy to see the different layers of the ground.[52] Yet, while making observations on the polar environment, Nordenskiöld took notice of the extinction of certain birds and animals, such as eiders, whales, walruses, and seals, because of intensive hunting.[53] Nordenskiöld perceived contemporary environmental problems and even proposed solutions, including hunting prohibitions or restrictions. David Wool has argued that we tend to think that awareness of the negative effects of humans on the environment arose only recently, but actually the ideas were there already 170 years ago when Lyell recognized the ecological phenomena and described them as clearly as any modern ecologist.[54] Richard Grove confirms this opinion by noting that from the late eighteenth century onward there were leading pioneers of environmentalism with a deep sense of historical perspective on environmental change and who also dared to warn of impending environmental crisis.[55]

Adolf Eric Nordenskiöld brought history to the blank North by reading its geology and traces of past life. His observations of geological processes and his interpretations of ancient environments influenced his opinions about nature. Fossils focused his thoughts on climate change. His watchful eyes led him to gather an extensive and remarkable fossil collection, which was then organized and processed by Professor Heer in Zurich, whereas before his scientific expeditions, only some scattered knowledge was available on Spitsbergen's geology. As a geologist and collector, Nordenskiöld opened up new fields for other scientists.

As the North is once again becoming the archive against which we measure global climate change, Nordenskiöld's experience has relevance. Through his hands-on experience in the North and fossil research, Nordenskiöld learned to understand the long geological processes moulding the surface of the earth. Rather than travelling to or through the spaces of the North like Richard Byrd's expeditions did, as described by Marionne Cronin in this volume, Nordenskiöld travelled through time. On his expeditions to the Arctic regions he encountered dramatic changes in nature, which made him aware of the vulnerability of the fragile northern environment. Writings on the Arctic fossils brought home by the Swedish expeditions opened up new ways of thought on the northern natural environment and created some of the first discussions of significant climatic changes that took place dozens of millions of years before our present time. By intently and intensely reading fossils, Nordenskiöld opened up new avenues for understanding global climate change phenomena and centred the scholarly gaze on the North as a record of those changes.

NOTES

1 Steven M. Stanley, *Earth and Life throughout Time*, 2nd ed. (New York: W.H. Freeman, 1989), 5, 9; see also Martin J.S. Rudwick, *The Meaning of Fossils: Episodes in the History of Palaeontology* (Chicago and London: University of Chicago Press, 1985).

2 For example, Margaret B. Davis, "Palynology and Environmental History during the Quaternary Period," *American Scientist* 57, 3 (1969): 317-32; Svante Björck and Per Möller, "Late Weichselian Environmental History in Southeastern Sweden during the Deglaciation of the Scandinavian Ice Sheet," *Quaternary Research* 28, 1 (1987): 1-37; G. Velle, J. Larsen, W. Eide, S.M. Peglar, and H.J.B. Birks, "Holocene Environmental History and Climate of Ratasjoen, a Low-Alpine Lake in South-Central Norway," *Journal of Paleolimnology* 33, 2 (2005): 129-53; L. Nevalainen, K. Van Damme, T.P. Luoto, and V.-P. Salonen, "Fossil Remains of an Unknown Alona Species (Chydoriadae, Aloninae) from a High Arctic Lake in Nordauslandet (Svalbard) in Relation to Glaciation and Holocene Environmental History," *Polar Biology* 35 (2012): 325-33.

3 J.M. Hulth, "Nordenskiölds bibliografi: Förteckning öfver A.E. Nordenskiölds skrifter," *Ymer* 2 (1902): 277-303. I use the name Spitsbergen instead of Svalbard because Nordenskiöld used the term. Nowadays the area is called Svalbard, and Spitsbergen is only a part of it.

4 A.G. Nathorst, "A.E. Nordenskiölds polarfärder och A.E. Nordenskiöld såsom geolog," *Ymer* 2 (1902): 209.

5 Only recently has the topic attracted more interest; for instance, in Finland, the John Nurminen Foundation has published two weighty books on the achievements of Nordenskiöld and navigation in the Arctic regions: Christoffer H. Ericsson, *Koillisväylä viikingeistä Nordenskiöldiin* (1992), and Matti Lainema and Juha Nurminen, *Ultima Thule: Arctic Explorations* (2001).

6 The Nordenskiöld Collection is stored in the National Library of Finland; it contains more than three thousand volumes of geographical and cartographical literature and travel books, plus a unique collection of about some twenty-four thousand historical maps. Since 1997, it has been in UNESCO's register, Memory of the World.

7 Nordenskiöld's reports and letters are available at the Center for History of Science at the Royal Swedish Academy of Sciences in Stockholm (hereafter CHS). The letters of his Finnish ancestors are filed in the National Library of Finland in Helsinki.

8 A.E. Nordenskiöld, *Nils Adolf Erik Nordenskiöld ur Svenskt Biografiskt Lexikon* (Stockholm: F&G Beijers Förlag, 1877), 3-4 (hereafter *Nordenskiöld*); A.E. Arppe, *Minnestal öfver Nils Gustaf Nordenskiöld* (Helsinki, 1866), 9-10.

9 Nordenskiöld, *Nordenskiöld*, 9.

10 Ibid., 10-12.

11 Ibid., 15.

12 Adolf Erik Nordenskiöld's letter to Nils Gustaf Nordenskiöld, 11 June 1856, National Library of Finland, Frugård collection 335 (hereafter NLFFc); David Wool, "Charles Lyell – 'the Father of Geology' – as a Forerunner of Modern Ecology," *OIKOS* 94, 3 (2001): 386.

13 George Kish, "Adolf Erik Nordenskiöld (1832-1901): Polar Explorer and Historian of Cartography," *Geographical Journal* 134, 4 (1968): 487-500.

14 Charles Lyell (14 November 1797-22 February 1875) was a British lawyer and the foremost geologist of his day. He is best known as the author of *Principles of Geology*.

15 Rudwick, *The Meaning of Fossils*, 201; Stanley, *Earth and Life*, 385-86, 433-34, 542; Wool, "Charles Lyell," 387.

16 Edouard Bard, "Greenhouse Effect and Ice Age: Historical Perspective," *C.R. Geoscience* 336 (2004): 618.

17 Adolf Nordenskiöld – Fr. W. von Otter, "Account of the Swedish North-Polar Expedition of 1868," *Proceedings of the Royal Geographical Society of London* 13, 3 (1868-69): 151-70.

18 Nordenskiöld's letter to his mother, Sofie Nordenskiöld, 29 August 1858, NLFFc. The original text that I have translated: "De vetenskapliga resultaterna hafva äfven såväl för Torell som mig varit ganska betydliga. Jag har funnit rikt petrofektförande lager af bergkalk formation (det v. säga nederste siktet från stenkolsperioden). Äfvensom yngre bildningar hvilka väl voro föga petrefakt förande men deremot kalksten eller brunkolslagad, till kisel förvandlat träd, vextaftryck, löfaftryck, lämningar af räffelser, frukter(?) ["?" is in Nordenskiöld's letter] o.s.v. fullständiga bevis på en för detta rik vegetation häruppe. Allt detta för Pappas räkning."

19 Nordenskiöld's letter to Nils Gustaf Nordenskiöld, 9 September 1858, NLFFc.

20 A.E. Nordenskiöld, *Redogörelse för sen svenska polarexpeditionen år, 1872-1873* (Stockholm: F.A. Norstedt & Söner, 1875), 109-10; A.E. Nordenskiöld, *Utkast till Spetsbergens geologi* (Stockholm, 1866), 33-34.

21 Nordenskiöld, *Spetsbergens geologi*, 34; Nordenskiöld – von Otter, "Account of the Swedish North-Polar Expedition of 1868," 155; A.E. Nordenskiöld, "On the Former Climate of the Polar Regions," *Geological Magazine* 2, 11 (1875): 525-32.

22 Nordenskiöld's letter to his wife, Anna Nordenskiöld, 29 August 1870, CHS, B01:3; Oswald Heer, *Die schwedischen Expeditionen zu Erforschung des hohen Nordens vom Jahr 1870 und 1872 auf 1873* (Zurich: Friedrich Schulthess, 1874), 18-19.

23 Alexander Leslie, *The Arctic Voyages of Adolf Erik Nordenskiöld* (London: Macmillan, 1879), 294, 373-74; A.E. Nordenskiöld, *Vegas färd kring Asien och Europa jemte en historisk återblick på föregående resor längs Gamla verldens nordkust: Förra delen* (Stockholm: F&G Beijers Förlag, 1881), 23; A.E. Nordenskiöld, "Vetenskapliga resultat af Nordenskiöldska Jenisei-expeditionen," *Stockholms Dagblad*, 26 October 1876.

24 Nordenskiöld, "On the Former Climate of the Polar Regions," 526-31; current knowledge holds that during the Cretaceous period continents moved toward their modern configuration. At the start of the period, the continents were tightly clustered, and Gondwanaland was prominent in the south. By the end of Cretaceous time, however, the Atlantic Ocean had widened, and Gondwanaland had fragmented

into most of its daughter continents (Stanley, *Earth and Life*, 385); A.E. Nordenskiöld, "Sketch of the Geology of the Ice Sound and Bell Sound, Spitzbergen," *Geological Magazine*, decade 2, vol. 3 (1876): 16-23, 63-75, 118-27, 255-67.

25 Nordenskiöld, *Spetsbergens geologi*, 33.

26 Adolf Nordenskiöld, *Den andra dicksonska expedition till Grönland* (Stockholm: F. & G. Beijers Förlag, 1883), 261-62.

27 Nordenskiöld's letter to his brother Otto Nordenskiöld, 25 October 1857, NLFFc.

28 Wilhelm Odelberg, "Adolf Erik Nordenskiölds samlingar och minnen från hans expeditioner," in *Nordostpassagen från vikingarna till Nordenskiöld*, ed. Nils-Erik Raurala (Helsinki: Helsinki University Library, John Nurminen Foundation, 1992), 191; Nordenskiöld's letter to his brother Otto Nordenskiöld from Hammerfest, 31 May 1858, NLFFc. In the letter, Nordenskiöld told Otto that he received money for his first expedition from the Swedish Academy of Sciences; *Nordisk familjebok 29: Tidsekvation – Trompe* (Stockholm: Nordisk familjeboks förlags aktiebolag, 1919), 374-76; Otto Torell demonstrated his opinions in his first work, *Bidrag till Spetsbergens molluskfauna jemte en allmän öfversigt av arktiska regionens naturförhållanden och forntida utbredning* (Stockholm: Typografiska Föreningens Boktryckeri, 1859).

29 Bard, "Greenhouse Effect and Ice Age," 603-38.

30 Charles Ricketts, "The Cause of the Glacial Period, with Reference to the British Isles," *Geological Magazine*, n.s. 2, 12 (1875): 573-80, 573.

31 F.W. Hutton, "Did the Cold of the Glacial Epoch Extend over the Southern Hemisphere?" *Geological Magazine*, n.s. 2, 12 (1875): 580-83.

32 Nordenskiöld, "On the Former Climate of the Polar Regions." The passage quoted by Professor Judd appears at p. 531, but the whole paper is well worthy of perusal, as is his paper "On the Geology of Icefjord and Bell Sound, Spitzbergen," *Geological Magazine* (1876): 16, 63, 118, 255. Nordenskiöld's "Expedition to Greenland" also appeared in *Geological Magazine* 9 (1872): 289, 355, 409, 516, and has material in it bearing on the former climate and extinct floras.

33 A.H. Green, "Professor Nordenskiöld on Recurrent Glacial Periods," *Geological Magazine*, n.s. 3, 1 (1877): 40.

34 A.E. Nordenskiöld, "The Nordenskjöld Greenland Expedition," *Nature* 29 (8 November 1883): 40.

35 Nordenskiöld, *Redogörelse för en expedition till Grönland 1870* (Stockholm: Norstedt, 1871), 998-99; A.E. Nordenskiöld, "The Nordenskjöld Greenland Expedition," *Nature* 29 (1 November 1883): 10-13, and *Nature* 29 (8 November 1883): 39-42, 40.

36 Nordenskiöld, "The Nordenskjöld Greenland Expedition," *Nature* 29 (8 November 1883): 40.

37 Joshua Lindahl, "Dr. N.O. Holsts's Studies in Glacial Geology," *American Naturalist* 22 (1888): 589-98, 705-13.

38 Ibid., 595.

39 Ibid., 597, emphasis in original.

40 Derek R. Mueller, Warwick F. Vincent, Wayne H. Pollard, and Christian H. Fritsen, "Glacial Cryoconite Ecosystems: A Bipolar Comparison of Algal Communities and Habitats," *Nova Hedwigia*, Suppl. 123 (2001): 173-97; Susanne E. Roeck, "Überlebenskünstler in Eis und Schnee," *Wissenswerk*, December 2009: 14-15; Mark Jenkins,

"True Colors: The Changing Face of Greenland," *National Geographic*, June 2010, http://ngm.nationalgeographic.com.

41 Mueller et. al., "Glacial Cryoconite Ecosystems," 174.

42 Ibid.; Roeck, "Überlebenskünstler in Eis und Schnee," 14-15.

43 Nordenskiöld, "On the Former Climate," 530; Stanley, *Earth and Life*, 166: More traditional geologists clung to the idea that large blocks of continental crust could not move over the surface of the earth. Thus, when the distribution of certain living and extinct animals and plants began to suggest former connections between presently separated land masses, most geologists tended to assume that great corridors of felsic rock (the most abundant material in continental crust) had once formed bridges that connected but later subsided to form portions of the modern seafloor. Today, it is recognized that this could not have occurred because felsic crust is of such low density that it cannot possibly sink into the mafic rocks that underlie the oceans.

44 Oswald Heer, *Flora Fossilis Arctica, die fossile Flora der Polarländer* (Zurich, 1868-83); Heer also described in his *Flora Fossils Arctica* petrifactions collected by other expeditions, such as the English Franklin expeditions from the northern most archipelago of America, and those of Steenstrup in Iceland.

45 Heer, *Die schwedischen Expeditionen*, 14-16, 21.

46 Ibid., 41-42, 44-45.

47 Letter from Charles Lyell to Nordenskiöld, 5 January 1867, CHS, E01:15; Nordenskiöld, *Nordenskiöld*, 21-22.

48 C. Cooper King, "Manual of Natural History, Geology, and Physics; and Instruction for the Arctic Expedition, 1875: Second Notice," Review, *Geological Magazine*, n.s. 2, 3 (1876): 172-78, quotation at 175-76.

49 A.G. Nathorst, "A.E. Nordenskiölds polarfärder," *Ymer* 2 (1902): 141-206.

50 A.G. Nathorst, "A.E. Nordenskiöld såsom geolog," *Ymer* 2 (1902): 207-24; Nordenskiöld, "Vetenskapliga resultat."

51 This and the preceding two quotations are from Ralph O'Connor, *Earth on Show: Fossils and the Poetics of Popular Science, 1802-1856* (Chicago: University of Chicago Press, 2008), 4.

52 Nordenskiöld, *Den andra dicksonska expedition*, 296; Nordenskiöld, *Om den under Professor A.E. Nordenskiölds ledning nyligen afgångna Svenska polarexpeditionen*, June 1868, CHS, F02a:7; Nordenskiöld, "Grönland 1870," 978; A.E. Nordenskiöld, *Svenska expeditionen till Spetsbergen år 1864 om bord på Axel Thordsen under ledning af A.E. Nordenskiöld* (Stockholm, 1867), 89; Nordenskiöld, *Vegas färd*, 5, 7-8, 23.

53 For example, Dunér et al., *Svenska expeditionen till Spetsbergen och Jan Mayen* (Stockholm: Norstedt and Söner, 1867), 45-47, 112-13; Nordenskiöld, *Vegas färd*, 148-49.

54 Wool, "Charles Lyell," 391.

55 Richard Grove, "Environmental History," in *New Perspectives on Historical Writing*, 2nd ed., ed. Peter Burke, 261-82 (University Park, PA: Pennsylvania State University Press, 2004), 262-63.

Technological Heroes
Images of the Arctic in the Age of Polar Aviation

MARIONNE CRONIN

On 10 May 1926, the *New York Times* excitedly announced that Richard Byrd, an American naval officer, had become the first person to reach the North Pole by air.[1] News of Byrd's successful flight sparked a wave of public celebration. On his return, Byrd and his expedition were met in New York by large crowds and feted with a tickertape parade whose snowstorm of paper briefly transformed New York streets into polar canyons, bringing, at least for a moment, the polar fantasy to the city's heart. As the newspaper with exclusive rights to cover Byrd's expedition, the *New York Times* was sure to make the most of this excitement, and soon the paper was full of stories celebrating Byrd's heroic flight (Map 3.1).[2]

This heroism took two distinct forms. In one type, Byrd's flight was hailed as a symbol of technology's transcendent progress – no longer would fur-clad explorers struggle for months across the ice, marshalling their dog teams or dragging their sleds toward the pole; now they soared gracefully above the treacherous ice, borne on the wings of modern progress. Simultaneously, other narratives firmly grounded Byrd in the polar environment, embedding him in an earlier tradition of heroic polar explorers whose personal courage allowed them to overcome the dangers of the polar wilderness. In these stories, the aircraft was present primarily as the heroic explorer's tool rather than the source of his power. In the tensions between these two narratives one can see the struggles of a culture seeking to integrate aviation, and modern technology more generally, into its image of polar exploration and the polar environment.

By changing his movement through and mediating his experience of the polar environment in particular ways, the aircraft altered the relationship between explorer and physical environment. Given this interaction's crucial role in previous heroic narratives, the aircraft's ability to lift the explorer above the Arctic landscape and to shield his body within a technological shell had potentially significant consequences for the aviator's status as a polar hero. Although his new-found aerial viewpoint and ease of travel promised omniscience and omnipotence, the pilot's seat in his cockpit far above the icy surface seemed to subvert the heroism of polar travel. And yet, both Byrd and his chroniclers sought to present his flight as part of the tradition of heroic polar exploration. To do so, they produced complex, sometimes ambivalent, reimaginings of the polar environment that sought to preserve a space for masculine heroism within a technologically advanced modern world.

This ambivalence about the place of aircraft in polar exploration articulated broader cultural anxieties about modern industrial society. Whereas articles celebrating the aircraft's triumph over the Arctic emphasized the technology's disciplining power to bring the untamed wilderness under the pilot's control and make it subject to his gaze, coverage of Byrd's flight that sought to present it as an heroic episode deployed images and tropes that signalled their participation in the wider historical phenomenon of antimodernism. Emerging in the later nineteenth century in response to the social, political, economic, and cultural changes that attended the rise of modern industrial societies, antimodernism articulated a suspicion of the enervating "overcivilization" promised by the luxuries of modern urban life and sought to redress this by turning to the more intense physical or spiritual experiences that could be found in other, more primitive places, cultures, or times.[3] The press coverage that sought to link Byrd to the heroic tradition of exploration did so by presenting the polar environment as a space of untouched wilderness – unruly, undisciplined, wild, and perilous. Tracing the way chroniclers of the Byrd expedition imagined the polar environment and the way these images underpinned their narratives illuminates the cultural tensions created by Byrd's use of aircraft in the polar Arctic. Challenging the assumption that the use of aircraft automatically produced a totalizing image of the conquered North, a study of this coverage shows us how writing the technology into narratives of polar exploration

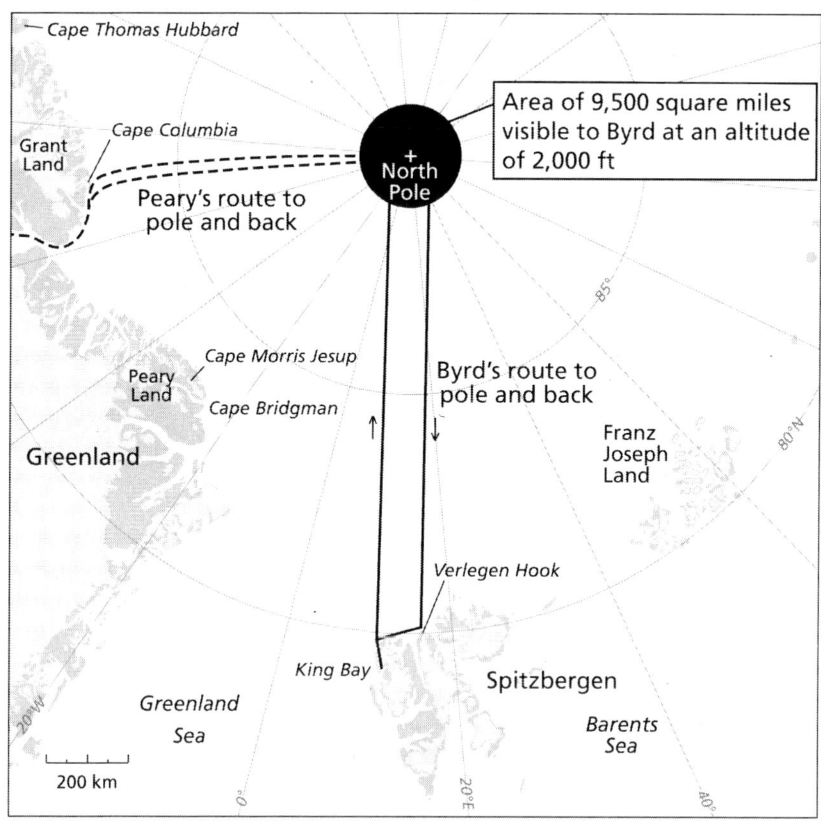

Map 3.1 Richard Byrd's flight to the pole showing his area of visibility. Based on "What Byrd Saw at the Pole," *New York Times*, 11 May 1926.

produced complex, layered, sometimes contradictory, and ambivalent images of the polar environment.

Technology and the Polar Landscape

To appreciate the challenge aircraft presented to polar exploration's heroic status requires a brief review of the environment's central role in nineteenth- and early-twentieth-century Anglo-American exploration narratives. In these narratives, the polar Arctic is constructed, first and foremost, as an othered space that exists outside the boundaries of the modern, civilized, everyday world. It is, instead, imagined as a space of primitive nature – wild and threatening, but simultaneously beautiful and pure. Sealed off by these dangers, the Arctic's secret mysteries can be penetrated only by the courageous explorer whose journey through this perilous place transforms him into a hero. This imaginary landscape, by turns beautiful and treacherous, provides the necessary backdrop for the explorer's enactment of narratives of heroism, masculinity, and national identity. It is the hazardous encounter between the explorer's physical body and the powerful forces of this untamed nature that provided the opportunity for him to demonstrate his heroism through his ability to preserve his character in the face of this malevolent, capricious nature. Indeed, his heroism did not depend on his physical survival but on his ability to display the right sort of character even in the face of death.[4] In these images, one can see clear echoes of the antimodernist turn to the primitive as a means of restoring modern masculinity.

One persistent theme in antimodernist discourse is the concern that the luxuries of modern urban life would emasculate men by replacing lives of physical struggle with lives of ease.[5] In response, advocates of what Anthony Rotundo calls passionate masculinity recommended physical activity as a means whereby men could access their primitive masculinity.[6] This need for physical activity was often tied to outdoor pursuits: to recover their true masculinity, men needed to escape the comforts of modern life by entering the wilderness. Free of the urban, men could shed the layers of feminizing conventions and recover their true characters through energetic activity. Although the advocated activities ranged in intensity from countryside cycling to wilderness travel, the aim was to prevent the decay of modern masculinity. In this context, wild, unspoiled nature – that is, spaces seemingly beyond the reach of the industrial urban world – provided the antidote to overcivilization. The poles, then, could be seen as the most remote and therefore the most pristine and wild of spaces, and polar exploration as the apotheosis of masculine wilderness pursuits.[7] When linked to the heroic quest narrative, this image created a powerful cultural narrative that bound together images of heroism, masculinity, and the polar wilderness.

Centred as it was in the male body, this heroic narrative depended on the direct engagement between the explorer's body and the physical environment. Within this paradigm, the aircraft's appearance was deeply destabilizing, as its use dramatically transformed the physical relationship between explorer and environment, removing him from direct contact with his physical surroundings by raising him above the struggles of the ice, shielding him from the polar environment within the plane's body, and moving him through the Arctic at speeds his predecessors could only dream of. Not only did the technology fundamentally reconfigure this relationship, it also threatened the polar Arctic's status as untouched wilderness. Aerial expeditions set technologies at their centre in a way that could not be ignored, and the juxtaposition of high technology and polar landscape seemed to strike at the roots of the heroic polar narrative. These machines breached the separation between the urban and the wild by shrinking the distance that seemed to divide them, penetrating the illusion that the Arctic existed as a space outside the modern world, placing a symbol of the modern technological world in the heart of the wilderness. Indeed, historians such as Max Jones have argued that the use of aircraft marked the end of the heroic age of polar exploration.[8] The story, however, was more complicated. In the coverage of Byrd's flight one can see writers reacting to the consequences of introducing aircraft into this environment. Some responded by celebrating the technology as the polar conqueror. Others sought to shore up existing concepts of polar heroism. But neither response was uncomplicated, and each had implications for how their authors conceptualized the polar environment.

Heroic Machines

As with previous expeditions, the process of narrating the Arctic began even before Byrd's expedition left New York, as Byrd and his supporters sought to present the flight as a worthwhile endeavour.[9] Byrd was not the first to make the case for using aircraft in the Arctic. Nineteenth-century explorers Salomon Andrée and Walter Wellman had both attempted, unsuccessfully, to use lighter-than-air craft on polar expeditions. In the twentieth century, the doyen of polar exploration himself, Roald Amundsen, had made an aerial attempt to the pole in 1925, and Byrd had been part of Colonel Donald MacMillan's 1925 US naval aerial expedition to Greenland. Despite this history, Byrd's flight required justification, and the rhetoric deployed in the run-up was designed to affirm its legitimacy.

As part of this project, expedition coverage in the months prior to Byrd's departure sought to construct the flight as a scientific expedition, an image that particularly appealed to Byrd's financial backers, particularly J.D. Rockefeller.[10] This coverage foregrounded the modernity of Byrd's equipment and methods, presenting them, and therefore the expedition, as cutting-edge and thoroughly modern.[11] In this context, the aircraft itself was presented as both an instrument for acquiring and the product of scientific knowledge. In outlining his plans, for example, Byrd himself grouped the plane with the expedition's other observational equipment, emphasizing its status as an instrument deployed by the scientific explorer.[12] Stories lovingly described the inner workings of both the instruments and the aircraft, underscoring its technical sophistication and its resulting ability to transcend the difficulties of the northern environment.[13] Byrd too was presented as a man of science. In particular, his technical knowledge of the aircraft and its performance, along with his contribution to the development of the expedition's navigational instruments, was offered as evidence of his scientific and technical expertise.[14]

Although previous terrestrial expeditions had sought to present themselves as scientific endeavours, to do so was particularly necessary in the context of aerial expeditions.[15] In the United States, aviation had a long association with public entertainment, and the development of barnstorming in the early 1920s only reinforced aviation's image as merely irresponsible stunting.[16] The "air-minded," like Byrd, who sought to present aviation as a worthwhile business, faced images of aviation that dismissed it as purposeless, reckless, and foolhardy. In order to undercut these images, advocates such as Byrd sought to represent it as sober, reliable, and useful; characterizing the flight as a scientific expedition was a means to this end.[17] To achieve this, Byrd emphasized the expedition's potential contributions to scientific knowledge of the Arctic, of the atmosphere, and of navigational methods and technologies. Moreover, Byrd argued that his flight to the pole would demonstrate the reliability of aircraft technology. Byrd's use of the expedition's utility as a means to argue for the legitimacy of commercial aviation offers an interesting contrast to previous explorers who had pointed to their scientific activities as a way of distancing themselves from any hint of practical interests that could tarnish their image as disinterested, pure heroes and the act of exploration as one that gained its legitimacy from its very inutility.[18]

Where the expedition's scientific contributions were an important element in legitimizing its pursuit in the months leading up to Byrd's flight, in the aftermath of his triumphant return, these narratives took on a more jingoistic tone, rejoicing in the triumph of American technology over the Arctic environment. Drawing upon concepts of aviation as a transcendent technology, technological triumph narratives celebrated the machine's power to shrink polar distances and to transcend the dangers inherent in previous methods of polar travel. They also celebrated the ability of the aerial gaze to penetrate the Arctic's secrets.

As Joseph Corn has pointed out, in early-twentieth-century America, the act of flight itself represented the ability to achieve mastery over the environment, to break free of the limits that bound human beings to the earth, and to conquer space and time.[19] In coverage of Byrd's flight, narratives of technological triumphalism not only argued that aircraft could transcend the Arctic's dangers, they went so far as to present aircraft as a technology that could tame the wild Arctic or, as Byrd put it, pull the Arctic's teeth.[20] One *New York Times* story commented, "The Arctic has succumbed to engineering." Indeed, the conquest was so complete, one writer postulated that even "Santa Claus probably drives a sedan."[21] To Russell Owen, the *New York Times* journalist who accompanied the expedition and provided almost daily updates by wireless telegraphy, Byrd's polar flight provided evidence of technology's power to bring even the most wild or hostile environment under the pilot's control, make it subject to his will, and conquer it.[22]

In particular, these stories used the contrast between Byrd's flight and Robert Peary's polar journey to emphasize the aircraft's tremendous power to overcome the Arctic. Where Peary's expedition had taken 429 days to struggle over jagged ice and pressure ridges, these stories trumpeted, Byrd and his co-pilot soared above these obstacles, leaping hundreds of miles to the pole and back in a matter of a mere hours.[23] The aircraft's power served to compress time and space, thereby bringing the Arctic's vast tracts within the explorer's and, by extension, the reader's grasp. With the support of modern technology, the previously ungovernable polar wilderness could be encompassed and, by implication, known and controlled.

In its ability to annihilate distance, the airplane laid the polar interior open to the explorer's gaze. At the same time, the explorer's new aerial vantage point reinforced the ability of his vision to render the North knowable. As Stephen Bocking argues, aerial views can be a powerful tool deployed to render remote landscapes legible, and therefore controllable.[24] In the context of Byrd's flight, this was expressed in the aerial explorer's ability to survey large amounts of unexplored territory in a very short space of time. Byrd himself reckoned that from his cockpit he could see fifty-five miles in every direction, which allowed him to survey an area approximately two thousand times larger than the five square miles Peary could have observed from the ground. This aerial platform, Byrd argued, allowed him to visually explore approximately 160,000 square miles during his relatively short flight.[25] In this narrative, the technology functioned as a scientific instrument that could penetrate into the Arctic's interior and compel it to reveal its secrets. Thus, the airplane could transform large swathes of territory from unknown to known, from wild to conquered, within a matter of hours. As Byrd expressed it, aircraft enabled men to "gaze upon [the Arctic's] charms and discover its secrets, out of reach of those sharp claws."[26] Through the use of aircraft, these authors argued, the polar Arctic was rendered both accessible and legible.

The technology's ability to tame the wild Arctic offered the possibility of transforming the polar environment from wilderness to conquered landscape. No longer separated from civilization, aircraft now linked the Arctic to the industrialized urban centre. As George Putnam put it, "the Arctic was now in our suburbs," and according to the technological progress narratives, it was now a hive of activity, on the verge of becoming an aerial highway.[27] Through the use of aircraft, these writers argued, the Arctic could be transformed from a frightening, uncontrollable, otherworldly space into a familiar landscape of suburbs and highways. There had been hints of this in some pre-expedition coverage, but this argument was most prominent in post-expedition coverage.[28] By penetrating the Arctic's secrets and shrinking distances, the aircraft would open the region as a transport route and give access to its land and resources.[29]

The Perilous Arctic

Although the stories of technological triumph seem to suggest that the introduction of aircraft resulted in a complete transformation of American cultural images of the polar Arctic, the resulting renegotiated images were not ones wherein a conquered, biddable landscape simply replaced a threatening wilderness. Indeed, for the celebration of technological triumph to have any meaning, the risks associated with the polar environment could not be entirely erased. These narratives navigated a conceptual tightrope: although technology offered the prospect of taming the Arctic, it needed to remain a simultaneously conquered and hostile environment. Thus, despite technology's promise, the polar Arctic remained necessarily treacherous. Those who sought to celebrate technology's ability to conquer the Arctic found it necessary to present the environment as simultaneously dangerous, though they focused on its dangers for the machine rather than for the explorer. As Fitzhugh Green of the *New York Times* described it, the region remained "shrouded six months of the year in Arctic night. Blizzards must howl over it. Its rocky shores can only be rugged and iceworn." Given these dangers, the Arctic was a space that offered "little promise for escape should disaster overtake the airship on the way."[30]

In particular, authors concentrated on the risks the "formidable" polar ice posed for aircraft. In his memoirs, Byrd described it as

> covered everywhere with snow and criss-crossed with pressure ridges like a crazy quilt ... Some of these frozen leads, probably older than the others, had a layer of snow over them and looked as flat as a table, but I knew them for dangerous sirens. They gave the appearance of affording an excellent landing, but they were very probably too thin to hold the plane, which, in landing, would crash through to destruction.[31]

This description presented the Arctic as a treacherous environment whose false appearances could lull the unwitting into a false sense of security, luring them to their doom. This sense of malevolent danger was echoed in articles that described pressure ridges, ranging in height from a few feet to over fifty or sixty feet, that would have destroyed any plane attempting to land among them.[32] In these accounts, although the aircraft's power allowed it to leap over these obstacles, perilous threats remained present in the Arctic and a direct encounter between aircraft and the polar surface could destroy the plane. In this way, these narratives attempted to reconcile the competing demands of triumphal technological progress narratives with the need to preserve the polar environment as a space of danger.

In addition to providing the risks necessary to maintaining the flight's heroic status, these dangers were harnessed into the story of technological advancement as authors constructed the polar surface as a technological testing ground – a sort of cold-weather laboratory.[33] Byrd and the reporters chronicling his expedition described how on the expedition's first test flight it was quickly discovered that when skis replaced its original wheels, the Fokker aircraft's undercarriage transferred the plane's weight to the outside edges of the skis, digging them into the snow and creating extensive resistance. Over the course of the first attempts to take off from the snow runway, the skis, which had been specially reinforced for their Arctic mission, were "broken like paper in the rough snow" and the landing gear was damaged.[34] Not only did the axels need to be straightened but the skis themselves required reinforcement in order to withstand the pounding they received from the fully loaded plane moving over the snow and ice.[35] The test flight episode, including the difficulties with the skis, functioned narratively in two ways. On the one hand, it presented the Arctic as an extreme environment with conditions beyond the limits of everyday experience that tested the technology to its limits. At the same time, Byrd and his colleagues' success in adapting the technology to suit its Arctic operating conditions provided evidence of American technological expertise's capacity to triumph over even the most challenging environment, and the technology's ability to operate in these conditions guaranteed its capacity to perform in any environment.

Just as authors deployed descriptions of the polar surface to underscore the dangers of polar flight, Byrd also recruited the polar atmosphere to support his expedition's technological significance. In describing the challenges of polar flying, Byrd pointed to his previous experience on the MacMillan Greenland expedition, arguing that the Arctic's unpredictable weather and air conditions presented terrible obstacles. As Byrd described it, poor weather, which appeared quickly, could mean a loss of visibility that hampered the pilot's ability to navigate and left him vulnerable. Turbulent air currents could also produce difficult flying conditions. Byrd described one flight on the MacMillan expedition as a great battle against the elements, the rough air tossing the plane about like a leaf. The pilot's survival depended on his ability to control the aircraft under these conditions. Even at rest the planes were vulnerable to gales that swamped anchored aircraft and drove icebergs among them.[36] To minimize these risks, Byrd had opted for a spring expedition that avoided the changeable summer weather but meant the expedition would encounter colder temperatures, which would present their own technical challenges. Byrd argued that the use of techniques developed for northern Canadian flying, combined with the scientific instruments and increasing knowledge about Arctic aviation, would enable his expedition to conquer the polar environment.[37] In this way, these difficult conditions would test the aircraft and its operators but would ultimately enable them to demonstrate their technical prowess.

These threats to the technology also translated into risks for the explorers. Should the technology fail, Byrd and his co-pilot, Floyd Bennett, would be marooned in this hostile environment. Despite having emergency supplies, neither was an experienced polar hand, and the two ill-prepared men would have struggled to survive. In his memoir, Byrd used a description of an oil leak on one of the plane's engines to highlight their vulnerability. When the leak was discovered, the plane was approximately seven hours from Spitsbergen and still one hour flying time from the pole. If the leak had persisted, Byrd and Bennett might have been stranded in a badly disabled machine, but they decided to press on despite the risk. Luckily, the leak was the product of a loose rivet and stopped once the oil level had fallen below the rivet.[38] Byrd used the episode as a means of demonstrating his heroic character, persisting in his quest even in the face of danger, but it also served to underscore the aviator's dependence on his technology and his vulnerability if forced to step outside the aircraft. With both engines and planes untested in prolonged cold-weather operation, and without the surface experience of an explorer like Amundsen, the crew placed its fate in the lap of a machine whose failure could leave them isolated and vulnerable on the pack ice.

No Longer a Space for Heroes?

These narratives sought to preserve the Arctic as a space of danger and risk, yet they did so in order to reinforce an overarching narrative of technological progress. In doing so, however, they presented a significant challenge to the explorer's heroic status at the centre of the story. Although the pilot retained some significance as the airplane's guide, in many of these stories the power to bring the Arctic under control rested with the machine. Without it, the aviator was presented as isolated and vulnerable to the Arctic dangers. Foregrounding the technology in this way was important for the air-minded narratives that sought to present the airplane as a heroic, reliable, and powerful machine.[39] The emphasis on science and technology also served to justify the flight by constructing it as a scientific expedition, both observing the Arctic environment and testing the technology in the Arctic laboratory. In addition, presenting the technology as the central figure allowed it to take on the role of national avatar, defining American identity as synonymous with technological progress and using an American technology as a means of projecting American power internationally, much as other expeditions had been used to project power and status. But doing so was deeply destabilizing for the heroic exploration narrative.

Of course, Byrd's flight was not the first polar expedition to deploy cutting-edge technology. Nineteenth-century expeditions, for instance, set out with a vast array of equipment that made the ships, in Francis Spufford's phrase, a veritable "Crystal Palace afloat."[40] Peary's polar expedition, likewise, had been supported by the latest in steam-powered ship technology, and even Robert F. Scott had attempted to use motorized vehicles. However, where the chroniclers of these expeditions had been able to gloss over the role of technology or to incorporate the machines into their narratives,[41] the power celebrated by the technological progress narratives surrounding Byrd's flight made it impossible to blend the technology into the background. By focusing on the machine, the coverage of Byrd's flight foregrounded the changing relationship between explorer and his physical environment. Given the centrality of this interaction to existing concepts of heroic exploration, the explorer's removal from the polar surface to the cockpit's interior threatened to destroy the explorer's heroic status. Indeed, for some, the deployment of advanced industrial technology meant these flights were no longer heroic ventures.[42] For these critics, the advantages offered by technology meant that aerial assaults on the pole were not quite "sporting."[43]

The aircraft's potential to undercut the explorer's heroism also fed into broader interwar anxieties about the consequences of technology for American masculinity. As we have seen, the emergence of industrial civilization brought with it antimodernist anxieties about the consequences of overcivilization and a reactionary desire for more authentic experiences that could be achieved through a retreat into untouched, primeval wilderness.[44] The pressures on physical notions of masculinity only increased through the early twentieth century as technological developments, such as the assembly line, amplified the routine and monotony of industrial work and as office work became increasingly feminized. Indeed, though still limited, growing female participation in public life seemed to some to further restrict the spaces available for authentic masculinity.[45] Wilderness experiences, whether hunting trips or Boy Scout camps, offered a means to escape the malaise of modernity and supplied an arena for men to reinvigorate their masculinity.

As we have also seen, wilderness exploration, and specifically polar exploration, provided an ideal type of this return to nature. The use of aircraft, however, seemed to strike at the very heart of this muscular masculinity, redefining the polar Arctic by making it an easily accessible zone of commercial activity. The technology also disengaged the explorer from the physical environment, protecting him from the risks, dangers, and struggles of polar travel and eliminating his body as a locus of physical agency.

Within this context it is easy to see why historians such as Max Jones have argued that the age of heroic exploration came to an end with the Antarctic voyages of Scott and Amundsen. However, the impulse to construct Byrd's polar flight as part of this tradition remained strong, perhaps because of the threat represented by the increasing pressure being exerted on masculine spaces. Given the aircraft's presence, these narratives could not simply rehearse existing tropes, but through judiciously deployed of images of the Arctic environment their authors were able to renegotiate the image of the heroic explorer and the polar environment to accommodate the new technology. The key strategy the authors used to resist this threat was to reinforce the image of the polar Arctic as a dangerous wilderness or primitive nature that menaced both the explorer and the plane.

In the months leading up to Byrd's flight, writers in the *New York Times* sought to link Byrd's expedition to the earlier tradition of heroic exploration by presenting the Arctic as a mysterious, unknown region full of potential dangers. In the words of Fitzhugh Green, who interestingly was also one of the authors to celebrate Byrd's flight as a technological triumph, the Arctic was "the last large blank space on the globe; the last really sizable geographical mystery left to man."[46] By outlining the large expanse of unexplored Arctic space, the maps used to illustrate these stories reinforced the Arctic's image as unknown and therefore outside man's control. In tracing Byrd's route, the maps depicted both his flight path and the route Peary followed to the pole.[47] This image served to link Byrd's expedition to the heroic tradition Peary represented. It also emphasized the Arctic's vast mystery by contrasting the swathes of unknown territory with the thin lines these expeditions traced. At the same time, however, these maps conveniently ignore the routes followed by nineteenth-century British and American expeditions, thus overstating the region's mystery. They also glossed over the history of failure connected to some of these expeditions, including John Franklin's disappearance and the traumatic experiences of expeditions led by Elisha Kane, Charles Hall, and Adolphus Greely. Nor do they acknowledge the presence of the people for whom parts of this region were home. Through these lacunae, these maps construct the Arctic as an empty, unknown wilderness that provides the setting for heroic American triumphs.

The sometimes florid text that accompanied the maps presents the region not only as mysterious but also as wild, bleak, and savage:

> The shores bordering on the polar sea are nothing save an utter waste of frost-gnarled rock and gravel abandoned by ages of ice and bitter cold.
> It is across this barren wilderness that these daring fliers will soon make their effort to clear up the globe's last great geographical mystery ...
> ... North of Canada spreads in confusion that group of ice-girt islands which have been the scene of so many hardships and deaths.[48]

As these comments suggest, a rearticulation of these heroic narratives required that the Arctic be more than empty: it must also be perilous. Only thus could it provide the necessary risks. In the lead-up to the flight, Russell Owen, the *New York Times* reporter assigned to the expedition, described the Arctic as "a country of raw elements where humans tread softly and with precautions against the silent menace of the North."[49] Even when commenting on the pleasant weather, Owen injected a sense of potential peril: "The

Arctic smiles now, but behind the silent hills is death."[50] Coverage also focused on the dangers these changeable conditions, particularly the fogs and spring winds of late May and June, would present for these aerial expeditions.[51] Advertisers too underlined the risks. A Mobiloil advertisement, for instance, highlighted the hazards presented by Arctic ice, fog, and winds – hazards that its engine oil would help mitigate.[52] These were the risks Byrd's expedition must confront, battle, and overcome.

Emphasizing the perils of aircraft travel and the technology's vulnerability to the physical conditions allowed writers to present the Arctic as a space of trial for both the men and their machines. By surviving and conquering the environment, the machines would prove their reliability and practicality, whereas the men would demonstrate their courage by venturing into this unknown, dangerous environment onboard a technology that made "the old style of discovery by stout surface ships and dog sledges seem ... to be safe by comparison."[53]

When Byrd returned, those who wished to celebrate his achievement framed it in terms of his ability to overcome these hazards. For example, the mayor of New York rhapsodized,

> How proud we are that on the perilous voyage through the air, braving unseen and unimagined danger and hardships, daring the unknown frozen wastes and the bitter winds whose icy breath spelled death, over seas of perpetual ice which no human eye had ever before beheld, you made your departure from this the harbor of New York.[54]

Byrd was a hero because he had "braved the bleak and bitter blasts of the icy North, [making his] painful and dangerous way across the trackless wastes" as he plunged toward a "terrible Unknown."[55] Byrd himself emphasized the size and power of the Arctic, describing it as a desolate vastness, a bleak and bitter landscape, and a region of death.[56] These images presented the Arctic as an environment where a wild, untamed nature threatened the explorer's survival – it was dangerous, unpredictable, and unfeeling.

Two episodes played key roles in these heroic narratives: the expedition's arrival at Kings Bay and the flight itself. When Byrd and the *Chantier* arrived at Kings Bay, they found the wharf taken up by another ship and, in order not to lose any time, decided to land the planes and supplies by ferrying them through the ice flows on jerry-rigged rafts. Coverage of this episode emphasized the risks, describing how the men "battled with the ice, wind and tide, ferrying everything across the half-mile of ice-jammed water on a frail raft built on four lifeboats."[57] Byrd would return to the theme in his later article for *National Geographic Magazine*, emphasizing the disassembled plane's narrow escapes from fatal damage on its journey from ship to shore.[58] It was depicted as an anxious time, since any misadventure, particularly with the plane, could have resulted in the premature end of the expedition. This episode provided an opportunity to highlight the technology's vulnerability to the Arctic's perils as, unlike in its fully assembled stage, the dismantled machine was at the mercy of the wind and ice. With the technology disabled, the incident also foregrounded the direct encounter between the expedition's bodies and the Arctic environment, as the crew's efforts to pilot the planes through the dangerous ice pans placed their bodies in direct contact with the environment and simultaneously allowed the crew to demonstrate their willingness to risk their lives in narrow escapes from the wind and ice.[59] Focusing on this episode enabled Byrd and reporters to draw attention to the physical male body in its struggle with the polar environment.

This particular narrative involved the recitation of another polar exploration trope: the cheerful crew. Where the central hero might be expected to display stoic endurance, the polar crew, often drawn from a lower social class than the hero, were expected to display a cheerful courage. Byrd's crew did not disappoint. In an article with the stirring title "Make Epic of the North," Owen described how the crew laboured with frostbitten fingers and frozen toes, chapped faces and running eyes, but a "happy, laughing disdain of cold, hunger and weariness."[60] When recounting the battle to land the aircraft, Byrd also stresses the crew's good humour:

> All knew the fate of the expedition was in their hands and that a single mishap could end all our plans and dreams. It was the most heartening thing imaginable to see them plunging ahead as if failure were impossible. Their laughing confidence in the race with serious dangers was a source of inspiration, encouragement and reward during those anxious hours ...
>
> Tired as all are, every man aboard and ashore wears a broad smile today. The fact is that the whole thing was done in what one might almost call the picnic spirit. Our fellows have a huge capacity for enjoying themselves under the most trying circumstances and their cheeriness was perhaps what carried us through, for there were many moments when gloomy natures would have given up in despair.[61]

Similarly, in his *National Geographic* article, Byrd underscored the crew's "enthusiastic industry and courage."[62] Byrd commented that the men displayed a cheerful bravery throughout these trials and the ones that followed as they readied the plane for the expedition. Running through these stories are also descriptions of the physical challenges presented by Arctic conditions, which allowed their authors to present the region as a parlous environment and to reaffirm the physical dimensions of aerial expeditions.

But what of the flight itself? It is in descriptions of these sixteen hours that the ambivalent character of the renegotiated Arctic environment becomes clearest. Although the triumphal narratives emphasized the transcendent experience of leaping over surface obstacles, other elements of the coverage reveal a much more complicated image. We have seen how even narratives of technological triumph left a space for the hazards the Arctic might present for the plane. The same was true in descriptions of the explorer's engagement with the Arctic. Removed from the surface, the explorer now experienced the cold and difficult weather through the plane. From this perch, his primary method of engaging with the Arctic surface was visual. This should suggest a powerful, totalizing, disembodied gaze, but it was not so clear-cut. As Anders Ekström argues, the aerial view can be both disembodied and profoundly physical, producing sensations of dizziness and vertigo.[63]

According to theories of the dominant aerial gaze, the pilot's seat high above the polar ice should produce a conquered, legible environment and, indeed, Byrd and his air-minded chroniclers celebrated the new power to survey the Arctic landscape offered by the aircraft. At the same time, however, Byrd described how this aerial vantage point brought home the utter scale of the Arctic ice; its vast sweep and its profound emptiness. Rather than sheering the Arctic of its mystery, this view accentuated the region's otherworldly qualities. The experience of flying over the vast ice made Byrd feel as if he and Bennet were

> insignificant specks of mortality flying there over that great, vast, white area in a small plane, speechless and deaf from the motors – just a dot in the centre of 10,000 square miles of visible desolation.
>
> We felt no larger than a pin point, as lonely as a tomb and as remote and detached as a star. Here in another world, far from herds of people, the passions and smallness of life fell from our shoulders.[64]

Even as it convinced him of the aircraft's ability to encompass the wilderness, the experience of flying through the Arctic made Byrd feel as if he had been taken outside himself.

Rather than simply confirming his ability to control the environment, Byrd found that the experience of Arctic flight could be profoundly disorienting. As Byrd put it:

> Time and direction became topsy-turvy at the Pole ... No matter how the wind strikes you at the North Pole it must be travelling north and however you turn your head you must be looking south and our job was to get back to the small island of Spitzbergen which lay somewhere south of us![65]

Moreover, geomagnetic phenomena at the poles meant that compass readings were inaccurate. Nor could pilots rely on navigation by sight over a constantly shifting sea of "featureless" ice.[66] While these difficulties were used to highlight the exceptional navigation skills that allowed Byrd to successfully negotiate the Arctic's perilous conditions, thereby reinforcing his heroic status, they also contributed to an image of the Arctic as a disconcerting, alien environment that operated outside the rules of the everyday world.

At the same time as it produced this transcendent, sometimes disorienting aerial view, the experience of flying through the Arctic remained profoundly physical. In describing his flight Byrd recalled the physical discomfort of remaining almost motionless for hours on end; the deafening, unending engine noise; and the frozen fingers gained when taking positional observations out the plane's trap door.[67] As with Byrd's navigational skills, these experiences were deployed to reinforce Byrd's heroic status as a man who could withstand the hardships of polar travel. Whereas technological triumph narratives sought to guarantee the plane's heroic status by pointing to the Arctic as a threatening environment, celebrations of Byrd's heroism sought to present the aircraft's interior itself as a difficult and sometimes perilous environment that mirrored the physical challenges of surface expedition. As in previous narratives, the explorer's ability to face this environment continued to function as a signifier of his character, allowing him to demonstrate his courage and resolution in the face of these hardships.[68] Accounts of the flight and of the struggle to unload the aircraft constructed the Arctic as an environment that would provide the necessary risks and challenges to test this heroic masculinity.

These narratives presented the Arctic as disorienting and sometimes treacherous, yet it remained simultaneously alluring, though its pull had a dark undertone. Both Byrd and the *New York Times* reporters depicted the landscape as hauntingly beautiful. Describing the scene when the fliers returned to Kings Bay, Russell Owen waxed lyrical:

> It was a scene never to be forgotten, for the northern light still shone as brilliantly bright and clear as when the plane left. On the white mountains the air was so crystal clear every crevice could be seen in the peaks of ice miles away. Hardly a cloud was in the sky, which was a beautiful pale blue. The snow and the ships resting on the mirrorlike surface of the water or tied to bergs seemed part of a heroic picture.[69]

But this preternaturally beautiful place was one that could easily turn on the explorer. As accounts of the risks posed by the polar ice and weather demonstrated, this image of the Arctic was not one of a conquered, pastoral landscape but of an unpredictable wilderness, one whose mysteries had lured men north for centuries – sometimes to their deaths.[70]

This image of the Arctic played a key role in securing the expedition's heroic status. By preserving its status as a wild space, these images of the environment reinforced concepts of the Arctic as a place in which men could continue to enact a more authentic masculinity. As part of this, the Arctic's perils created the opportunity for Byrd to demonstrate his heroic character by pitting himself against an extreme and hazardous environment. The Arctic's presentation as an otherworldly space also contributed to the heroic narrative's rearticulation. Depicting the region as an exceptional environment meant chroniclers could present acts performed in this environment as similarly exceptional. Moving through this otherworldly landscape could transform the aviator into a hero as he entered a space where others dared not go. The polar aviator likewise becomes a heroic figure, as Tom Crouch argues, as his ability to soar above the world separates him from the mere mortals that inhabit the terrestrial sphere.[71] By emphasizing the dangers, these narratives placed the pilot in the role of the lone pioneer who took himself beyond the known limits of his machine in order to challenge a hostile and unpredictable wild nature. By facing these risks, Byrd could demonstrate his heroic masculine character.[72] This heroism was both physical, illustrated by his ability to endure physical privation, and personal, evidenced by the strength of character that allowed him to pursue his goal even in the face of mortal danger and to ultimately triumph, thereby conquering the natural world. Thus, this image of the Arctic allowed Byrd to fit into pre-existing heroic narratives.

An Environment in Transition

The tensions in the ways in which writers imagined the environment through which Byrd flew reflect the ambivalences of a culture in the process of renegotiating its images of both heroic masculinity and the polar environment in the context of technological modernity and the anxieties of antimodernism. On the one hand, aircraft seemed to promise the power of a transcendent method of travel that could subdue and render legible an unruly nature. Conversely, this promise was deeply destabilizing for notions of heroic masculinity bound up in exploration narratives. Likewise, the view from the cockpit proved both powerful and disorienting, while the experience of flying could not erase the physicality of Arctic travel entirely. Nor did the Arctic become a completely subdued environment, since accounts of both technological triumph and heroic exploration continued to rely on images of the Arctic as unpredictable and dangerous.

These tensions left a space wherein their authors could begin to construct an image of the explorer as technological hero. Indeed, in particular articulations of the narrative it is precisely embracing the technology that enables the man to be a hero. As Roger Bilstein and Joseph Corn argue, part of the complex of cultural meanings that clustered round the airplane in the 1920s was the image of flight as a thing of wonder and grace that could lift one above the industrial, urban landscape and transcend the petty ugliness of modern life.[73] Here was a technology that offered a bastion of individualism in the age of mass mechanization. Indeed, it is precisely the aircraft's capacity to enable the explorer to escape civilization and to venture into the untamed spaces that allows him to demonstrate his heroism.[74] At the same time, writing the machine into the Arctic narrative provided a symbolic taming of this wilderness. In this context, the polar aviator could be cast as a daring lone pioneer who escaped the limits of society and challenged and conquered a hostile nature. Thus, the *New York Times* could rejoice in Byrd's accomplishment as a demonstration that "despite the security and luxuries of modern life, men can still match the heroism and endurance of other days."[75] As Byrd himself perceptively noted in his memoir, the hero offered an avatar for the nation, and through the consumption of these narratives, the readers could vicariously share in the reassertion of masculinity offered by the image of the polar aviator.[76]

For all that Byrd seemed to offer a hero for the nation, it was still a limited definition of who the hero could be. Not only was it a highly gendered image but the polar hero remained solidly white and middle or upper class. Indeed, Byrd's expedition offered a simpler heroic narrative than Peary's, as Byrd's flight was accomplished without relying on the female and Inuit labour that had to be written out of Peary's narrative.[77] In many ways, Byrd offered a similar image of American Yankee masculinity to that constructed around

Charles Lindberg the following year, and as Charles Ponce de Leon comments, this reserved middle-class masculinity provided a powerful counterpoint to the flamboyant, swaggering, urban, working-class, and often immigrant masculinity enacted by other celebrities of the 1920s, such as Jack Dempsey and Babe Ruth.[78] As in previous polar expeditions, the supporting cast of masculine labour was conveniently encompassed by its characterization as the loyal band of cheerful heroes depicted in the struggle to land the expedition at Kings Bay. Thus, like other polar narratives, the depiction of Byrd's heroism constructed a classed, racialized, and gendered image of the polar hero.

While asserting Byrd's heroic status, these narratives did so in particular ways that reflected the context and preoccupations of interwar America and the impact of introducing aircraft into the very heart of the practice of exploration. The tensions in these narratives manifested the contradictory pulls between the desire to present these flights as evidence of technological triumph and the desire to assert a space for masculine heroism within the context of prevailing anxieties about the status of masculinity in a highly technological world. They also illustrated the friction between the air-minded desire to present these flights as evidence of aviation's reliability and practicality and the demands of the heroic narrative. These tensions could have pulled the narrative apart. However, the deployment of specific images of the Arctic environment enabled the narratives to incorporate the aircraft into the polar landscape and to construct an idea of masculine heroism that not only neutralized the threat presented by the machine but also offered an image of the technological hero.

These images present the polar Arctic as an environment in transition. In the interwar period, the North American states, supported by new technologies (particularly aircraft), had begun expanding into the North with the MacMillan expedition and similar Canadian government-sponsored expeditions along the Arctic coasts. In the interior, economic and political frontiers moved north, transforming the sub-Arctic environment as they did so.[79] In the 1920s and 1930s, the Arctic was not yet the scientific-military space it would become in the Cold War era, but it was no longer the imaginatively isolated setting of the age of heroic exploration.[80] The use of aircraft had changed that, making explicit and unignorable the ties that had always connected the Arctic and the rest of the world.

Placing the technology squarely at the centre of their narratives, Byrd's expedition and the other aerial expeditions of the period demanded that narratives negotiate the implications of modern technology for how American cultures of exploration and masculinity understood the Arctic environment. The narrative images pulled in contradictory directions. The Arctic is at once the conquered, disciplined environment connected to the metropolitan environment through the aircraft, and simultaneously it is a space of unruly, wild nature that supports the heroic narrative. In the inter-war period, the polar Arctic remained the space of heroic explorers, if only just. But one can see the growing intensity of its construction as a disciplined space, an image that would underlie postwar ideas of the militarized, scientific, and administered Arctic.

NOTES

1 "Byrd Flies to North Pole and Back," *New York Times*, 10 May 1926. Although later some have challenged the accuracy of Byrd's claim, at the time, his observations were reviewed and confirmed by a panel of scientists convened by the National Geographic Society and his success was generally accepted.

2 These exclusive arrangements were an important way for expeditions to raise the funds that supported their adventures. As Beau Riffenburgh points out, coverage of polar exploration was also an important part of competition for readers and explorers, and newspapers had developed an interdependent relationship through the later nineteenth and early twentieth century. Newspapers helped construct the explorer as a heroic public figure and helped fund his expedition, while explorers' exploits provided the fodder for dramatic stories that attracted readers. For the historian, however, these exclusive arrangements do mean that immediate press coverage of an expedition is often limited to a single source. Although other newspapers might report on Byrd's expedition, they were often reduced to repeating *New York Times* stories. As a result, the following analysis draws heavily on the *New York Times.* Beau Riffenburgh, *The Myth of the Explorer: The Press, Sensationalism, and Geographical Discovery* (London: Belhaven Press, 1993).

3 See T.J. Jackson Lears, *No Place of Grace: Antimodernism and the Transformation of American Culture, 1880-1920* (New York: Pantheon Books, 1981). For an examination of antimodernism in the North, see Christina Adcock, "Many Tiny Traces: Antimodern Anxieties and Colonial Intimacies in the Canadian North," Northern Environmental History Workshop, Trent University, Peterborough, ON, October 2011.

4 For analyses of the cultural history of Anglo-American polar exploration, see Lisa Bloom, *Gender on Ice: American Ideologies of Polar Expeditions* (Minneapolis: University of Minnesota Press, 1993); Jen Hill, *White Horizon: The Arctic in the Nineteenth-Century British Imagination* (Albany: State University of New York Press, 2008); Max Jones, *The Last Great Quest: Captain Scott's Antarctic Sacrifice* (Oxford: Oxford University Press, 2003); Sarah Moss, *The Frozen Ship: The Histories and Tales of Polar Exploration* (New York: Blue Bridge, 2006); Michael F. Robinson,

The Coldest Crucible: Arctic Exploration and American Culture (Chicago: University of Chicago Press, 2006); and Francis Spufford, *I May Be Some Time: Ice and the English Imagination* (London: Faber and Faber, 1996).

5 Michael S. Kimmel, *Manhood in America: A Cultural History*, 2nd ed. (Oxford: Oxford University Press, 2006); E. Anthony Rotundo, *American Manhood: Transformations in Masculinity from the Revolution to the Modern Era* (New York: Basic Books, 1993).

6 Rotundo, *American Manhood*, esp. 222-30.

7 Mountaineering provided a similar expression of physical masculinity. For an examination of the interaction between mountaineering, masculinity, and scientific practice, see Bruce Hevly, "The Heroic Science of Glacier Motion," *Osiris* 11 (1996): 66-80.

8 Jones, *The Last Great Quest*, 13, 290-92.

9 For an examination of the mobilization of cultural resources that sits behind exploration, see Felix Driver, *Geography Militant: Cultures of Exploration and Empire* (Oxford: Blackwell, 2001), 8.

10 "Byrd to Explore Arctic from Air," *New York Times*, 31 January 1926; "Rockefeller Gifts Take a New Trend," *New York Times*, 1 February 1926; "North Pole by Airplane," *New York Times*, 26 March 1926.

11 "Byrd Sails North this Month to Hop 400 Miles to Pole," *New York Times*, 1 March 1926; "Byrd Planned Trip to Arctic for Years," *New York Times*, 4 April 1926.

12 Richard E. Byrd, "Byrd Outlines Plan to Reach the Pole," *New York Times*, 28 March 1926. Compare this strategy with the resistance experienced by the Danish scientist Lauge Koch in advocating for the use of aerial surveying as a scientific technique: Christopher J. Reis, "Lauge Koch and the Mapping of North East Greenland: Tradition and Modernity in Danish Arctic Research, 1920-1940," in Bravo and Sörlin, *Narrating the Arctic* (Canton, MA: Science History Publications USA, 2002), 199-234. See also Robert A. Nye, "Medicine and Science as Masculine 'Fields of Honor,'" *Osiris* 12 (1997): 60-79.

13 "Byrd Sails North this Month to Hop 400 Miles to Pole," *New York Times*, 1 March 1926; "Polar Flier to Use a Fokker Plane," *New York Times*, 7 March 1926; "The Fokker Three-Engined Airliner to Be Used by the Byrd Arctic Expedition," *New York Times*, 5 April 1926.

14 "Byrd Says Wilkins May Still Go On," *New York Times*, 21 March 1926.

15 See Driver's exploration of the developing relationship between exploration and scientific practice in *Geography Militant*, especially 38-43 and 50-67. For an examination of the relationship between science and images of the Arctic see Bravo and Sörlin, *Narrating the Arctic*.

16 Dominick A. Pisano, "The Greatest Show Not on Earth: The Confrontation between Utility and Entertainment in Aviation," in *The Airplane in American Culture*, ed. Dominick A. Pisano, 39-74 (Ann Arbor: University of Michigan Press, 2003), 39-42, 51.

17 "Byrd Planned Trip to Arctic for Years," *New York Times*, 4 April 1926; "Byrd Hops Off for the North Pole: Shifts Plans, Makes Direct Dash and Hopes to Be Back in 24 Hours," *New York Times*, 9 May 1926.

18 Robinson, *The Coldest Crucible*, 31-54.
19 Joseph J. Corn, *The Winged Gospel: America's Romance with Aviation, 1900-1950* (New York: Oxford University Press, 1983); Tom D. Crouch, "'The Surly Bonds of Earth': Images of the Landscape in the Work of Some Aviator/Authors, 1910-1969," in Pisano, *The Airplane in American Culture*, 201-18.
20 Richard Byrd, "Byrd Describes Flight Back from Pole," *New York Times*, 17 May 1926.
21 Russell D. Owen, "Norge Landing Crew Arrives on Gunboat at Kings Bay Base," *New York Times*, 26 April 1926.
22 Russell D. Owen, "Making Arctic History," *New York Times*, 11 May 1926.
23 "Peary's Pole Trip Ate Up 429 Days," *New York Times*, 10 May 1926; "Flier's Wife Is Jubilant," *New York Times*, 10 May 1926.
24 Stephen Bocking, "A Disciplined Geography: Aviation, Science, and the Cold War in Northern Canada, 1945-1960," *Technology and Culture* 50, 2 (2009): 265-90.
25 "What Byrd Saw at the Pole," *New York Times*, 11 May 1926; Richard Byrd, "Byrd Describes Flight Back from Pole," *New York Times*, 17 May 1926.
26 Richard E. Byrd, *Skyward: Man's Mastery of the Air* (New York: G.P. Putnam's Sons, 1928), 168. See also p. 176.
27 "Putnam in Arctic and Wife Here Talk," *New York Times*, 25 August 1926; "Amundsen Departs: Rails at Overcoats," *New York Times*, 7 March 1926; "Weather Bureau Aids Byrd Polar Flight," *New York Times*, 25 March 1926; "Spitzbergen Rings with Arctic Gaiety," *New York Times*, 28 April 1926; "From Greenland's Icy Mountains and Elsewhere," *New York Times*, 16 May 1926; "Would Put Our Flag Over the South Pole," *New York Times*, 24 June 1926; "Byrd Favors Start from Jones Sound," *New York Times*, 3 September 1926.
28 "This World," *The Youth's Companion*, 25 March 1926, 12.
29 "Polar Stories by Radio," *New York Times*, 16 March 1926; "Byrd Knocks Down His Plane to Ship It," *New York Times*, 1 April 1926; "Northward Ho!" *London Independent*, 116 (13 March 1926), 3954; Fitzhugh Green, "Massed Attack on Polar Region Begins Soon," *New York Times*, 7 March 1926.
30 Fitzhugh Green, "Massed Attack on Polar Region Begins Soon," *New York Times*, 7 March 1926.
31 Richard Byrd, "Byrd Filmed the Top of the World," *New York Times*, 16 May 1926.
32 William Bird, "Byrd Hops Off for the North Pole," *New York Times*, 9 May 1926.
33 For examinations of the Arctic as a scientific space, see Stephen Bocking, "Science and Spaces in the Northern Environment," *Environmental History* 12, 4 (2007): 867-94, Matthew Farish, "The Lab and the Land: Overcoming the Arctic in Cold War Alaska," *Isis* 104, 1 (2013): 1-29; Matthias Heymann et al., "Exploring Greenland: Science and Technology in Cold War Settings," *Scientia Canadensis* 33, 2 (2010): 11-42, and Bravo and Sörlin, *Narrating the Arctic*. See also Bruce Hevly's exploration of the role of danger as a guarantee of scientific authority, "The Heroic Science of Glacier Motion," *Osiris* 11 (1996): 66-86.
34 R.E. Byrd, *Skyward*, 159-60.
35 William Bird, "Byrd Tests Skiis [sic] on Fokker Plane," *New York Times*, 4 May 1926.
36 R.E. Byrd, *Skyward*, 137-38, 142, and 144.

37 "Byrd Ship Ready to Sail Tomorrow," *New York Times*, 4 April 1926; Richard Byrd, "Byrd Outlines Plane to Reach the Pole," *New York Times*, 28 March 1926.
38 R.E. Byrd, *Skyward*, 177-79.
39 Richard Byrd, "'Off for Pole with a Zoom'; Byrd Describes His Hop-Off," *New York Times*, 15 May 1926.
40 Spufford, *I May Be Some Time*, 160.
41 See, for example, Robinson's analysis of Peary's strategies for incorporating his ship into his narrative of primitive polar travel: *The Coldest Crucible*, 126-27.
42 "Peary's Aid Praises Both Polar Airmen," *New York Times*, 14 May 1926; "Real Arctic Honors," *New York Times*, 14 June 1926.
43 Tina Loo notes a similar anxiety about the use of technology in the male wilderness pursuit of sport hunting. Tina Loo, "Of Moose and Men: Hunting for Masculinities in British Columbia, 1880-1939," *Western Historical Quarterly* 32, 3 (2001): 296-319.
44 Rotundo, *American Manhood*, esp. 222-30.
45 Kimmel, *Manhood in America*, 127-33.
46 Fitzhugh Green, "Massed Attack on Polar Region Begins Soon," *New York Times*, 7 March 1926.
47 "The Polar Expeditions," *New York Times*, 11 April 1926.
48 "The Polar Regions, Scene of Hunt for the Unknown," *New York Times*, 11 April 1926.
49 Russell D. Owen, "Norge Now Housed in Kings Bay Shed: Bore Severe Test," *New York Times*, 8 May 1926.
50 Russell D. Owen, "Amundsen Inspects Base for the Norge," *New York Times*, 23 April 1926.
51 "Troubles of the Explorers," *New York Times*, 4 May 1926; William Bird, "Byrd Hops Off for the North Pole," *New York Times*, 9 May 1926.
52 "To the Top of the World – with Byrd," *New York Times*, 2 May 1926.
53 "The Polar Expeditions," *New York Times*, 11 April 1926.
54 "City's Greeting to Flier in Form of Scroll Declares His Exploit Honors New York," *New York Times*, 24 June 1926. See also "Honor Medal for Byrd Asked in House Bill with Advancement for Him and Pilot Bennett," *New York Times*, 22 May 1926.
55 "Thousand Acclaim Byrd at Luncheon," *New York Times*, 24 June 1926; Edwin Markham, "To the Top of the World," *New York Times*, 26 June 1926.
56 "Early News to Ft Worth," *New York Times*, 13 May 1926; "Byrd Describes Flight Back from Pole," *New York Times*, 17 May 1926; "City's Greeting to Flier in Form of Scroll Declares His Exploit Honors New York," *New York Times*, 24 June 1926; "Thousand Acclaim Byrd at Luncheon," *New York Times*, 24 June 1926; "Byrd Nearly Took to Fliers in Stocks," *New York Times*, 3 July 1926; "Amundsen Arrives with Polar Mates; Gets Noisy Greeting," *New York Times*, 4 July 1926.
57 Richard E. Byrd, "Two Byrd Planes Nearly Assembled; First Hazard Over," *New York Times*, 3 May 1926. See also William Bird, "Byrd Lands Planes, Braving Ice Hazard," *New York Times*, 2 May 1926.
58 Richard E. Byrd, "The First Flight to the North Pole," *National Geographic*, September 1926, 359-60.
59 R.E. Byrd, *Skyward*, 158-59.
60 Russell D. Owen, "Make Epic of the North," *New York Times*, 3 May 1926.

61 Richard E. Byrd, "Two Byrd Planes Nearly Assembled; First Hazard Over," *New York Times*, 3 May 1926.
62 R.E. Byrd, "The First Flight to the North Pole," 360.
63 Anders Ekström, "Seeing from Above: A Particular History of the General Observer," *Nineteenth-Century Contexts* 31 (2009): 184-207.
64 Richard Byrd, "Byrd Describes Flight Back from Pole," *New York Times*, 17 May 1926.
65 R.E. Byrd, *Skyward*, 178.
66 Richard Byrd, "Byrd Outlines Plan to Reach the Pole," *New York Times*, 28 March 1926; "Byrd Sails North this Month to Hop 400 Miles to Pole," *New York Times*, 1 March 1926; "Byrd Planned Trip to Arctic for Years," *New York Times*, 4 April 1926.
67 William Bird, "Peary's Observations Are Confirmed," *New York Times*, 10 May 1926; "Byrd Describes Flight Back from Pole," *New York Times*, 17 May 1926.
68 "A Great Achievement," *New York Times*, 10 May 1926; "After Reaching the Pole," *New York Times*, 11 May 1926; "Guggenheim Cables Praise," *New York Times*, 12 May 1926.
69 Russell D. Owen, untitled, *New York Times*, 11 May 1926.
70 Richard Byrd, "Byrd Describes the Obstacles Met on Reaching Kings Bay," *New York Times*, 14 May 1926.
71 Crouch, "'The Surly Bonds of Earth,'" 205.
72 It was an image of the heroic technological pioneer that would shortly be redeployed in depictions of another American aviator, Charles Lindberg: Dominick A. Pisano and F. Robert van der Linden, *Charles Lindbergh and the Spirit of St. Louis* (New York: Smithsonian Institution and Harry N. Adams, 2002).
73 Roger Bilstein, "The Airplane and the American Experience," in Pisano, *The Airplane in American Culture*, 16-35; and Corn, *The Winged Gospel*.
74 "A Great Achievement," *New York Times*, 10 May 1926.
75 "The Times and Exploration," *New York Times*, 19 September 1926.
76 R.E. Byrd, *Skyward*, 194; "City's Greeting to Flier," *New York Times*, 24 June 1926; "Capital Give Ovation," *New York Times*, 24 June 1926; Richard Byrd, "Byrd Tells Why Men Explore Polar Regions," *New York Times*, 27 June 1926.
77 Bloom, *Gender on Ice*, 23-24.
78 Charles L. Ponce de Leon, "The Man Nobody Knows: Charles A. Lindbergh and the Culture of Celebrity," in Pisano, *The Airplane in American Culture*, 75-101.
79 See, for example, Liza Piper, *The Industrial Transformation of Subarctic Canada* (Vancouver: UBC Press, 2009); John Sandlos, *Hunters at the Margin: Native People and Wildlife Conservation in the Northwest Territories* (Vancouver: UBC Press, 2007).
80 On the Arctic in the Cold War era, see Bocking, "A Disciplined Geography"; Bocking, "Science and Spaces"; Farish, "The Lab and the Land"; Heymann et al., "Exploring Greenland."

PART 2

COLONIZING THE NORTH

Mounds, Middens, and Social Landscapes
Viking-Norse Settlement of the North Atlantic, c. AD 850-1250

JANE HARRISON

The Vikings of the North Atlantic, as vividly portrayed in film, art, and literature, epitomize a certain view of people of the North: rugged survivors, predatory, alarmingly charismatic, and shaped by their challenging climate and surroundings. Despite the development of a more rounded and subtle picture of their trading, craft, and farming activity, there remains an underlying, still potent presumption that Viking-Norse people were in thrall to, and even culturally limited by, their northern environments. As a consequence, even thoughtful commentators and researchers – especially those outside the North – have been slow to appreciate the social and cultural aspects of northern settlement landscapes. In this chapter I argue that these environments are extremely important in reaching an understanding of Viking appropriation of the landscape. In the context of this volume, the term "settlement landscape" embraces all the changes affected in the landscape by people in constructing domestic and work buildings, and in exploiting the surrounding natural resources. This created environment, however, was not just an indication of successful economic strategies: the form and character of the resulting broad settlement landscape expressed and reinforced social structures.[1]

In the North Atlantic, the very challenges posed by the natural environment fostered a particularly close relationship between people and landscape; this chapter explains the workings of that relationship over a period of around four hundred years. The building of these northern environments happened over generations, maintained by daily, everyday actions but becoming apparent over long decades. Dwellers in the northern regions over this period, such as the Vikings, acquired and honed a necessarily intimate knowledge of the potential and the problems offered by their natural surroundings. Environments were created by applying indigenous and imported knowledge to generate landscape-specific building technologies. This understanding not only supported day-to-day survival but was essential to shaping settlement landscapes rich in social meaning. Thus, the settlement landscapes of coastal regions moulded by Viking and Late Norse people in the North Atlantic provide an excellent example of the interaction of society, economy, technology, and nature to create distinctive cultural environments. The chapter illustrates this by focusing on the Northern Isles of northern Scotland and in particular on the west coast of the island of Mainland, Orkney (Map 4.1).

Viking-Norse communities of the ninth to thirteenth centuries exploited the environmental possibilities of northern coasts to create sophisticated social settings. These settlement landscapes encompassed environments altered by the use of a range of technologies to produce settings that embodied and reinforced local relationships of power and community. Building these environments in the Northern Isles relied on the long-rooted technological and environmental experience of indigenous and incoming groups. These landscapes of the North made considerable demands on the people living there, but with every turn of the spade and cast of the hook, Viking-Norse communities were also building social symbolism into their surroundings. They understood and exploited the landscape and its resources with a thoroughness that ensured economic survival and provided the wherewithal to shape and manipulate the landscape, creating a cultural environment. Thus, Viking landscape design was sustained by their environmentally intelligent appreciation of the coast and seascape. Archeological evidence from those Northern Isles of the United Kingdom reveals this dynamic and creative relationship between people and the environment, and demonstrates that the Vikings were not, in a straightforward way, the product of their natural environment.

Coastal Settlement Mounds of the Orkney Islands, Northern Scotland

One particular form of Norse settlement in the Northern Isles of Scotland is an excellent example of a created cultural landscape: deliberately built coastal settlement mounds. This chapter focuses on the landscape archaeology of

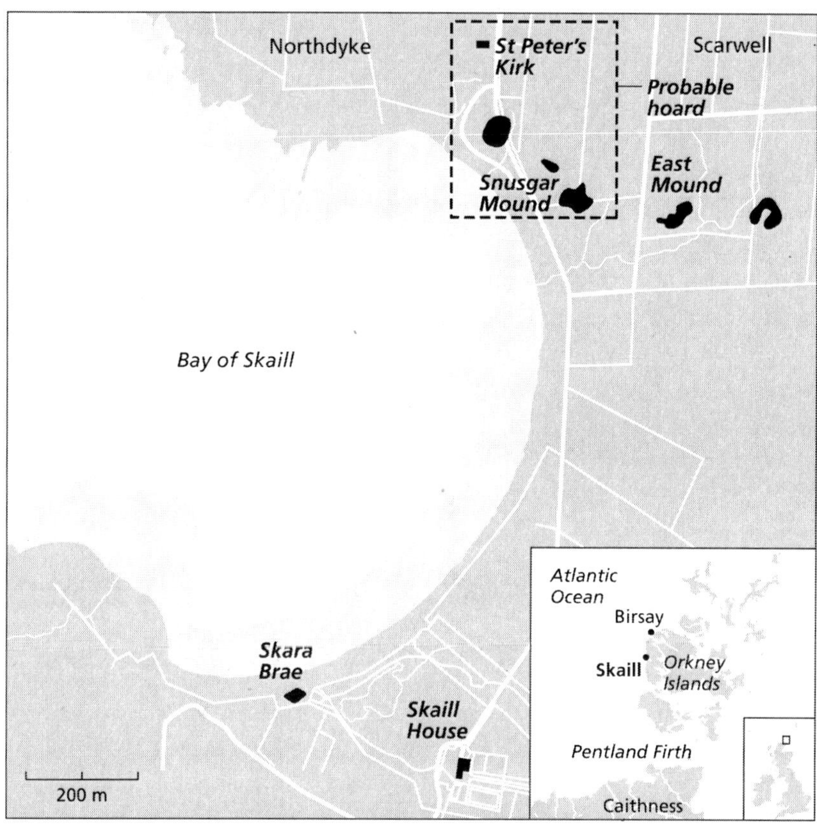

Map 4.1 The Bay of Skaill, West Mainland, Orkney. Inset shows location of the Birsay-Skaill Landscape Archaeology Project work.

the many Viking-Late Norse settlement mound sites found on the sandy bays of the Northern Isles of Orkney and Shetland.[2] These coastal settlement mounds are a defining element in the archaeology of the Orkney Islands in particular, but they have not been researched as a group across the islands, relatively few have been excavated, and fewer still investigated as part of a wider social landscape. Many earlier investigations concentrated on pre-Viking Iron Age broch mounds (a broch is a stone-built tower),[3] although a few Viking-Norse Age mounded settlements have been known for some time, in particular Jarlshof in Shetland.[4] More recent archaeological work has revealed that the prominent mounds on sandy bays frequently included settlements of the Viking-Late Norse period, found on sites including Birsay Bay, Skaill in Deerness, and the Bay of Skaill on the island of Mainland; Westness on Rousay; Tuquoy on Westray; Pool on Sanday; and Old Scatness and Underhoull/Hamar, on Shetland.[5]

The Orkney Islands exerted an understandable pull on Viking-Norse groups, whatever their original motivation for leaving the coasts of Norway (and perhaps Denmark). The archipelago was strategically located: a trading and raiding staging post on the seaways between Scandinavia, the north-east coast of Scotland, the east coast of England, the Western Isles, Ireland and the Irish Sea zone, and the further flung islands of the Faroe Islands and Iceland.[6] For those from Scandinavia seeking to settle, the Orkney Islands were equally attractive, offering the same environmental range as the coast and islands of Norway but with a generally more equable climate emphatically influenced by the Gulf Stream. Without the dramatic mountains and inlets of the coast of Norway, the Orkney Islands' low rolling hills and many gently curving, sheltered bays provided easy access to lowland and higher ground grazing, land for crops and sites for farmsteads close to the shore and sea. Transport between settlements could be quickly achieved by small boat, looping from bay to bay. The underlying geology – predominantly sandstone flags – produced outcrops of excellent building stone; the dominant coastal drift geology of windblown sand fostered light, easily worked soils, readily fertilized with turfs from nearby hills and seaweed from the adjacent shore. The growing season was longer than in many of the homeland areas, allowing the growing of barley in particular, along with flax and some oats; from the evidence of the byre sizes, cattle may have been overwintered outdoors. Although the winds buffeting the islands prevented the growth of anything other than small scrublike trees – willow and ash, for example – larger trees were available from the mainland of Scotland. Seen from further south, this was a challenging environment: windswept, cool, dark in winter, and lacking in deep, fertile soils; but to northern farmers and fishers, skilful handlers of boats, the Orkney Islands were lands rich in potential.

The Mound Farmsteads

The most significant settlements in the Northern Isles, on the prime sites offered by these sandy bays, were scattered farmsteads located on relatively low but visually dominant mounds (Figure 4.1). These farms were often small complexes of buildings clustered around a main hall, developed from a simpler, early longhouse with separate outbuildings. As in much of coastal Norway, settlement was scattered, with no villages, though more than one

Figure 4.1 The mounds at the Bay of Skaill, West Mainland, Orkney, under excavation, looking northwest. Although the mounds are relatively low, they are positioned so that they dominate the surrounding bay and the sea approach to the west. The Snusgar mound is on the left, the mound with the well-preserved longhouse is on the right, and St. Peter's Kirk is in the background. Photo by the author.

farmstead/hall cluster has been found in close quarters, for example at Underhoull, Shetland, and the Bay of Skaill, Mainland, Orkney. The settlement mounds are characteristically found on the windblown, sandy outer curves of the bays, sited to dominate sea approaches, with good views across both the shore and the grazing and farmland of the bay. With no readily available timber for construction, buildings were made of stone and turf, the imported timber or driftwood reserved for crucial roof-support posts. The basic longhouse arrangement is familiar across northwestern Europe from the Neolithic period onward: narrow, relatively long (ranging from less than ten metres to a maximum of thirty metres in the Northern Isles), furnished with side benches and a central hearth. Over time in the Viking-Late Norse period, paved courtyards, entrance approaches, and annexes were added, and houses were lengthened with the addition of extra rooms or integral byres. Later Norse farmsteads were thus the product of much amendment and change, giving them an unplanned and rambling appearance.

Of most interest to our purposes, these sand-mound settlements in Orkney and Shetland (and indeed in the Western Isles and on the northeast coast of Scotland) were also distinguished by a very specific building technology, with a range of related construction techniques. This technology enabled settlements to develop over one another on exactly the same spot. Earlier structures were extended, old walls reused, and changes made by building up over a very limited area. It was this propensity to stay put, and to superimpose, to build up over old yards, middens (rubbish heaps), surrounding ground surfaces, and buildings, that gradually augmented the mounds themselves. The drifting and piling of windblown sand contributed to that process. This concerted effort to stay in one place characterized settlements at the Bay of Skaill; at Pool, Sanday; at Skaill, Deerness; and other sites. Frequently, the first Viking buildings on a site were constructed reusing pre-Viking Pictish wall lengths and foundations, for example at Pool, Sanday; Buckquoy, Birsay; and Old Scatness, Shetland.[7] Viking structures were also erected by digging down into Neolithic middens, by infilling Late Iron Age buildings, and by reusing Late Iron Age wall lengths. This was not accidental. No limits on space necessitated clinging so assiduously to one spot. No physical constraints dictated this particular approach to building technology: that is the mounding of structures, paths, walls, and yards over one another, and the limiting of new buildings by referencing older arrangements. Considerable effort was expended in achieving these results. A mound developed by reusing older buildings and their related spaces was an end in itself – a cultural statement. Why that was and how it was achieved is discussed below.

Reidar Bertelsen and Raymond Lamb were the first archaeologists to extensively record and discuss the Viking-Norse aspects of settlement mounds in the Northern Isles of the United Kingdom.[8] However, they reported the mounds as confined to the northern islands of the Orkney archipelago: Sanday and North Ronaldsay. All subsequent work has demonstrated that coastal settlement mounds are a widespread phenomenon – either solely Viking-Late Norse or, more often, multi-period beginning with earlier settlement and capped with Norse settlement. Therefore, these mounds demand exploration as a significant, even defining, element in the settlement pattern of the Viking period in many areas of the North Atlantic.

In Norway, in the immediately pre-Viking period, the most significant mounds in the landscape would have been burial mounds, whereas the emphasis in Orkney was very much on settlement rather than burial mounds. A focus on the significance of settlement must have been more relevant for settlers coming to a land where the ancestors under burial mounds were not theirs. As is discussed later, the development of a mound symbolism focused on settlement grew out of an interaction between Norwegian incomers and the existing culture. The classic farm (settlement) mounds of northern Norway seem often to originate and flourish slightly later in the Viking Age, in the eleventh and later centuries, but perhaps there is a link.[9] Although there is unlikely to be a single, unnuanced North Atlantic-wide explanation for the creation of settlement mounds, the accumulation of some of those later Norwegian farm mounds may have been stimulated by the gradual development from the ninth century of the Orkney phenomenon. The Norwegian mounds – like the settlement of the Faroe Islands and Iceland – cannot be covered in any detail in this chapter, but many aspects of the discussion are relevant to Viking-Late Norse coastal settlement across the whole North Atlantic region. The cultural inspirations, and associated environmental technology, shaping socially structured settlement may indeed have travelled not just from east to west but also back again.

The mounds themselves, therefore, acted as cultural reference points linking the Northern Isles into a wider North Atlantic sphere. They looked back to the use of burial mounds in Iron Age Norway as territorial and political markers, and echoed the use of mounds as site locations for things, regional law-making, and administrative assemblies.[10] Rather than the accidental by-product of an economic system, settlement mounds in Orkney and Shetland were deliberately constructed to harness such associations. Taken across the sea and translated into the Orcadian context, mound symbolism was tied more closely into settlement because power in a newly settled but contested landscape related more nearly to land rights and resources. Mounded settlement was the product of the social and political intentions of the early Viking settlers in the ninth century and was then nurtured by subsequent generations from the tenth century into the twelfth century as the political and social framework of the Orkney Earldom developed. What follows therefore explores the powerful social messages evoked by settlement mounds. And, as the construction of the settlement mounds was reliant on the creative exploitation by fisher-farmers of the Northern Isles of the sea- and coast-scape, the full impact of the mounds' presence in the constructed landscape can only be understood against an appreciation of the nature of the Viking-Norse economy.

The Viking-Norse Economy in the Northern Isles

The broadly based, well-developed, and opportunistic economy of the Viking-Late Norse period was integral to the development of the cultural landscapes of Orkney's sandy bays. Environmental evidence from sites such as those at Birsay Bay and the Bay of Skaill (West Mainland, Orkney); at Pool, Sanday; and at Freswick in Caithness demonstrates how full use was made of all the resources of land, shore, and sea.[11] Crops were grown in the light soils that developed on windblown sand, in particular bere barley and oats. Flax, a crop introduced by the Viking settlers, was also cultivated and was particularly labour-intensive to harvest and exploit. The farmers kept cattle and sheep, as well as pigs, and although there is some evidence of a certain degree of specialization or difference in emphasis – for example, more dairying – on the whole, the farms kept their economies broad based as the most secure strategy.[12] The communal aspects of the economy have been rather neglected by commentators, perhaps because the settlement was relatively scattered but, as will become apparent, collaboration was essential to economic success.

This land-based strategy was complemented by garnering from the sea and foreshore. The farmers were skilled fishermen, and the shell-sand of sites like the Bay of Skaill has preserved many fish bones, mainly of the cod family – the bones of fish caught off the rocks and close to shore, as well as fished from bigger boats far out to sea. This offshore fishing in larger craft implies a level of cooperation between farmsteads. Later in the period, some sites provide evidence of specialization in producing dried stockfish and even fish-liver oil, but this economic development accompanied the wider cultural and political change of the twelfth century.[13] Excavations of the middens around farms and in the longhouses themselves have also unearthed plentiful seashells: some of clams, oysters, mussels, and razor clams, but most usually limpets and periwinkles. The collection of these two latter, now little regarded, shellfish has been held up as an example of gastronomic desperation – "starvation food" – but a swift trawl through nineteenth-century recipe books reveals that limpets and periwinkles held an honourable place in cooking, and were particularly prized for invalids.[14] Fisher-farmers also exploited the coast, harvesting seabird eggs and trapping the birds. Gannet bones from the Bay of Skaill all demonstrated the same leg-bone break, suggesting systematic trapping; eggshell was a surprisingly frequent find on the same site.[15]

The shore was essential for other resources: driftwood and seaweed – driftwood to supplement the timber imported from mainland Scotland and Norway for building in virtually treeless Orkney and Shetland, and seaweed for many purposes. Seaweed is an interesting component of the economy and perhaps of settlement symbolism. This plant belongs to both land and sea and evokes lives lived seamlessly from the shore out across the water.

Abundant on the rocky foreshore, seaweed provided a useful fertilizer for the sandy soils, supplementing the more traditional additions of peat and byre cleanings of hay mixed with dung. Thus, the weed of the sea added value to the soils of the land. Seaweed also augmented the cattle's grazing and was burned as a fuel, with the resulting ash being used to create floors in both byres and domestic areas. Turf was used extensively, especially in building. Walls with internal drystone facing were jacketed with turf; soil and turf were packed as a stabilizing core between double-faced walls. In areas of Orkney where peat was not plentiful, for example Sanday and parts of Mainland, turf was also burned as a fuel.[16] It would have been invaluable in the creation of charcoal. The Bay of Skaill site has provided ample evidence of the exploitation of then-more-plentiful scrub trees – in particular willow – for making charcoal.[17] Nothing of use in the landscape was neglected, and as is explained below, even the by-products of that usage – of the growing, fishing, cooking, spreading, burning – were crucial to the building technologies employed by the groups generating them.

Farmstead economies have been referred to as "self-sufficient," but in the relatively well-populated Northern Isles, this is a misrepresentation. They were, rather, non-specialist, but necessarily linked to other similar farms. Local cooperation in coastal areas maximized economic well-being: collaboration helped with harvesting, in maintaining flock and herd health, in crewing boats for fishing and longer expeditions, and in building farms. All of the more important settlement mound farmsteads yield objects such as combs, pottery, and metalwork from outside the Orkneys, signalling some participation in wider exchange or trade circles.[18] Furthermore, raiding and trading contributed vitally to the economy, and both demanded collective planning and action. The dominant mound farmstead provided a central location for deciding concerted action, and its carefully constructed physically prominent location also represented and monumentalized that cooperation. However, like the thing sites it echoed, it was far from an egalitarian symbol: the farmstead on the mound housed a locally powerful family in a context of wider social stratification. Local collaboration was, therefore, embedded in wider ties of obligation and tribute.

The Orkney Earldom had been established by the King of Norway toward the end of the ninth century and, in a relatively small, easily travelled archipelago, the earls and their powerful retainers made their presence felt as the earldom's political structures developed. Probably more than in the homelands, the demands and activities of the earls had a regular impact on most of the population of the islands. The vivid Orkneyinga saga (written in

Figure 4.2 The Bay of Skaill, looking south. This is a typical Orkney sandy bay, with the settlement mounds excavated by the Birsay-Skaill Project in the middle and left centre; St. Peter's Kirk is at the lower right-hand corner. Photo by the author.

thirteenth-century Iceland and mainly covering events of the tenth century onward) makes it clear that various forms of obligatory tribute-in-kind were a crucial part of the economy of the earldom.[19] All the farms in a locality must have needed to contribute something to meeting the requirements of the peripatetic earls and their retinues: food, formalized hospitality, and tribute. Organization through the earls' local representatives or leading families would have been necessary to ensure that every farm both played its part and received the compensation of patronage and protection. The mounds were, therefore, where local collaboration met political power; their construction was a physical expression of that social structure.

The Bay of Skaill, Mainland, Orkney: A Case Study

These northern settlement landscapes are still visible on the western mainland of Orkney, and in particular at the Bay of Skaill, where the University of Oxford's Birsay-Skaill Landscape Archaeology Project worked from 2004 to 2011 (Figure 4.2).[20] The project has been able to explore both well-preserved stone-built Viking-Norse buildings on settlement mounds and their immediate and wider landscape surroundings. The Bay of Skaill is best known for the extraordinary Neolithic village at Skara Brae, on the southern loop of the sandy bay. The Birsay-Skaill Project excavated two mounds with Viking-Norse archaeology on the northern side of the bay, drawing together the existing but rather fragmentary archaeological work on that period in West Mainland in the context of a new, extensive landscape survey. The survey work revealed that about fifteen sand mounds in the bay contained archaeological remains; two were excavated as part of the project. Work on the summit of the mound known as Snusgar discovered the badly damaged remains of a large ninth- to at least tenth-century longhouse.[21] Spreading out from the building were deep and carefully built outside surfaces and midden layers (containing some combination of domestic, fishing, and agricultural debris, with old turf, byre cleanings, and ash), which yielded excellently preserved environmental evidence. These accumulated layers, rich in organic material, constructed much of the notable bulk of the sand mound, giving height and creating conspicuous curves. Radiocarbon dates and datable objects from that mound confirmed dates beginning in the late ninth century and running through to the early twelfth (for the midden),[22] with one earlier seventh-century date for a horizontal layer lower down in the mound. This old land surface suggests that the core of the sand mound – as at Pool, Sanday, and other mound sites – is pre-Viking.

The second mound, to the southeast, had no name but contained the very-well-preserved remains of a considerable longhouse complex (Figure 4.3).[23] The first Viking-Norse structures were a classic bow-sided longhouse accompanied to the south by a smaller separate building at right angles to the main building; earlier buildings or middens seem to have provided the foundations for the smaller building, and the initial Viking construction took place into an already existing and visually well-placed, if relatively low, mound. A byre end with a flagged central passageway, stalls, and working areas was added to the longhouse, creating, by Northern Isles standards, a long structure of about twenty-five metres. The original house end, with its long side benches, hearths, and stone furniture, was also subsequently lengthened to create a substantial longhouse of nearly thirty metres. This final annex addition to the house was used predominantly for large-scale cooking, perhaps for the preparation of feasts. These changes and additions happened over a few generations, during which time substantial middens were also purposefully accumulated. All of the walls demonstrate repeated rebuilding: rather than start afresh, old features and wall lengths

Figure 4.3 A longhouse at the Bay of Skaill, looking east. The small building is right top, right of the flagged area. The byre end is in the background, the "hall" in the foreground, and the narrow cooking annex at the bottom. Photo by the author.

were incorporated, reskinned, shored up, and reused. Sometimes, as around the central paved yard south of the longhouse, this could be achieved only by complex combinations of stone infilling and wall refacing with new bracing upright stones (orthostats). Very particular stone-building technologies were being deployed. Yet, no effort was made to disguise these alterations to, or incorporations into, the stone walls; indeed, the cooking annex seems to be intentionally misaligned on the existing domestic area, with a stone from the older wall left protruding to mark the original alignment. This very deliberate referencing of older structures seems to have marked generational change, perhaps the passing of power from one local leader to another.

The smaller Viking building to the south, probably a byre or barn, was only the first in a complex sequence on that location. Another rectangular building was superimposed on top of gradually accumulated, and then deliberately and quickly infilled, layers within the first building, but rotated to run parallel to the longhouse. This second building reused several courses of the original walls in the north and the east. Then, fairly soon after (in archaeological terms – perhaps after a human generation), that building in turn was lengthened and realigned. Repeated building up of the walls over earlier foundations and stone courses on a slightly different line caused some stretches of wall to sag, requiring shoring, so on this occasion the builders had taken the trouble to straighten walls with whole new stretches of stone reskinning and refacing encasing the older walls. By this stage, the gradual tell-like accumulation of buildings, infilling, and floors over some hundred years in this part of the site was such that a flight of steps had to be added down from one of the exits of the small building toward the paved area in front of the longhouse's main door, to the north. Then finally, another generation later, the rectangular building was shortened to create a small square building. This well-flagged structure was probably still being used as an agricultural building in the twelfth century, though it might originally have functioned as a small bathhouse. It survived into the later twelfth century. By then, the entire length of the main longhouse had been downgraded to a byre and continued to be used as an animal shelter after its roof had collapsed.

Some years before, and around the time the second southern building was constructed, iron smelting and metalworking had begun in a sheltered, floored yard to its west. This metalworking area provided evidence for the

Figure 4.4 Deep layers of midden on the Castle of Snusgar mound, Bay of Skaill: stones, ash, and domestic and agricultural debris were carefully mounded over deep windblown sand. Photo by the author.

smithing of not just iron but also copper alloy, and would have been used at the time the longhouse to the north was at the zenith of its life, in the eleventh century.[24] Interestingly, a small hearth in the core of the hall had also been used for working copper alloy; this activity must have been conducted as an occasional demonstration of status, as intensive metalworking would have been unsustainable in the close, roofed environment of the narrow hall. In its heyday, therefore, the longhouse with its ancillary buildings was not a struggling farm but a building of some status. Several strands of evidence reinforce the idea that this farmstead, which during the tenth century encompassed the building on the Snusgar mound to the west, played a leading role in the local area: the imported pottery and jewellery, the considerable size and communal character of the hall building, the wealth of environmental evidence, and the presence of a metalworking area of some sophistication. Finally, Scotland's largest Viking-Norse silver hoard, dated to the period AD 950-70, was found very near the Snusgar mound in the nineteenth century.

Mound-Building Technologies and the Use of "Rubbish"

This broad approach to building technology, as seen at the Bay of Skaill mounds and typical of coastal mounded settlement sites, was driven by the desire to stay on the same sand-mound spot and demanded certain techniques. Structures, surfaces, and middens were built and layered up using material gathered from the local landscape. Although not usually recommended as a foundation for building, sand contributed to the creation of mounds, piling easily and mixing well with organic additions such as decaying turf, ash, and old straw. But sand also represented a complicating factor, with its fluid-like properties and tendency to shift and blow. Thus, superimposing yards, gardens, and buildings on sand mounds demanded very particular construction techniques beyond those methods applied to the wall courses. Buildings were dug into slopes and into the middens of previous settlement, as at Pool, Sanday. Existing, established wall lengths were reused, and buildings superimposed on more stable, organic-rich accumulated surfaces and infilled structures at sites at Birsay Bay, including Buckquoy; at Pool, Sanday; and at sites in Shetland, including Jarlshof and Sandwick. Mounds being eroded and cut into by the sea, like that at Marwick Bay or Skaill, Rousay, reveal the layer-cake accumulation of the deliberate mounding (Figure 4.4).[25] At all the sites, the effect was either to create a mound or, in the majority of cases, to enhance an existing visually dominant and economically well-placed natural or archaeological eminence. There were no obvious spatial limits to compel this "mounding," yet it seems to have been a very deliberate technological approach of some cultural significance in Viking-Norse settlement in the North Atlantic. And, as all the mounded sites occupied prime locations for pursuing the sort of economy described above, this was the chosen settlement construction approach of the more fortunate.

The mound-building process depended completely on the extensive use of organic resources harvested and nurtured from sea and coast: the resources available as a consequence of the broad-based coastal economy. Organics – ash, old turf, domestic and byre detritus, as well as the by-products of agriculture, fishing, and wild harvesting – occupied a critical place in mound dwellers' building technology and techniques. Thus, settlement style and economy were crucially linked.

First, however, it is important to consider briefly the significance and meaning of middens and rubbish beyond the obvious uses to which accumulated organic material could be put as fertilizers in agriculture. "Midden" is a term widely used to describe a spread or pile of domestic and/or farm waste, with the implication that the material was being collected for another use. However, in many cases, archaeologists give relatively little explanation of the intended use of middens other than as a source of fertilizer. The term "rubbish" seems to need no explanation but is probably anachronistic – very little in the pre-medieval era was discarded without care, or was without value. Archaeologists of the Iron Age claim there was no such thing as rubbish in that period.[26] Iron Age "rubbish" pits repeatedly yield examples of carefully chosen and laid objects and layers, demonstrating that even the apparent discarding of material after its final use or reuse was far from casual. Putting particular items into the ground and the disposal in a particular way of the organic remains of social occasions, agricultural productivity, or fishing success had cultural, probably ritual, significance. In Iron Age England, the often spectacular mounds resulting from the repeated piling over several generations of the debris of large seasonal, communal feasts amassed as growing testimony to significant events and to the capability of groups to put on a show.[27] Ideas of the worthlessness of, and lack of meaning attributed to, "rubbish" need also to be discarded when investigating the Viking Age. Stored refuse certainly had practical usefulness, but its accumulation and reuse also carried social meaning. Ash, for example, helped firm sandy floors and fertilize fields, but its spreading might also represent the hearth and the capability to keep fires burning. The rapid creation of middens suggested fecundity, agricultural capacity, and the accumulation of food-preparation debris from rich feasts. Almost everything organic in the Viking-Norse period had a continuing life of utility and, in application, were crucial elements in creating a landscape rich in social symbolism.

In the Viking-Norse Orkney Islands, "rubbish" and midden were crucial to the practical process of building repeatedly on the same location in a sand mound. Organic matter was stored routinely in various places and in various combinations for later use. The careful accumulation of middens has been found on every sufficiently investigated settlement site. On the Pool, Sanday, site, a Viking-Norse midden was carefully surrounded by a wall and paving to create a central settlement feature.[28] In the Bay of Skaill excavations, an ash heap was discovered carefully sealed with clay – presumably to stop it being blown or eroded away – but marked by spade cuts where material had been deliberately removed.[29] Work in 2011 discovered more of these clay-capped ash middens to the east of the longhouse settlement.[30] The uses of these stockpiled resources of ash, byre waste, seaweed, and turf extended beyond the relatively well-documented purpose of fertilizer for light sandy soils. One advantage to excavating in windblown sand is that anything *not* sand must have been added to the shell-sand–drift geology, by people, animals, birds, or the wind. The challenge is deciding how and why the additions were made.[31]

All the sites discussed have evidence for the introduction of ash, domestic waste, and so on to sand or light sandy soils for a number of purposes, including and beyond their use as fertilizer.[32] Domestic debris was mixed with sandy soil to create a firm inner core for drystone walling in structures; turf was used as an outer jacket for internally faced walls. The Pool, Sanday, and Bay of Skaill sites all provide excellent examples of the technique of digging down into firm, rich middens to provide support and insulation for new semi-sunken structures or extensions.[33] A less dense midden-enriched material was spread to level and – crucially – to stabilize sandy mound surfaces, enabling the further rebuilding or expansion of existing structures. The construction of firm yards or working areas and the provision of pathways depended completely on this technique, which can be seen especially well at Pool, Sanday; Jarlshof; and south of the longhouse at the Bay of Skaill. On the Snusgar mound, carefully structured midden spreads of domestic and byre waste, ash, and stone surfaces established layered, firm foundations for building up the sand mound. In working or access areas around the buildings, organic material such as food-preparation waste, crop processing, and fishing debris seems to have been intentionally discarded and spread in the knowledge that each addition to the sand helped reduce erosion and sand-blow.

Ash was especially important and is consistently found spread on house floors, in byres, and on outside working areas. Charcoal and ash from the turf-peat fires in the main Bay of Skaill longhouse were concentrated around the hearths but also dispersed across the whole floor.[34] The greasy, grey clay-like residues of cooking fashioned especially firm surfaces: in the metal-working area south of the hall, charcoal rake-out from successive hearths had been mixed with this material to form a laminated sequence of floors.

Thus, the drive to stay on a particular spot and to superimpose in sand relied on the creative application of a carefully chosen range of organic materials, supporting a particular range of construction technologies. These materials were available as a result of the coastal dwellers' thorough exploitation of the surrounding landscape and seascape. Living on a mound that gradually increased over generations reflected economic success and announced the mound dwellers' ability to control both the resources of the surrounding landscape and the people of that area. Maintaining and increasing a substantial mound signalled access to a considerable workforce for building and other labour, and to the accumulated debris of the frequent feasts, well-stocked byres and barns, and successful fishing and foraging expeditions so crucial to the physical construction of the settlement site. But mound living did not merely proclaim economic dominance and control of people; the mound *itself* had considerable social significance.

The Symbolism of Mounds

The first Viking people to settle on Orkney arrived to find an already inhabited and well-exploited landscape.[35] The new arrivals needed to legitimize their claim to control land and to establish social prominence. Different tactics were called for than in virgin territories such as the Faroe Islands and Iceland. Building over and with existing structures sent a powerful message: a degree of continuity with previously important communities was being signalled, while the takeover of land was demonstrated beyond doubt. Almost without exception, Viking-Norse settlement builds over or in very close proximity to Pictish/Late Iron Age dwellings. The more-high-status Viking settlements, such as at Birsay, West Mainland, were built near more important Iron Age sites. In an essentially non-literate society, visual symbols were important, and power structures needed to be inscribed on the surrounding environment, on the settlement landscape. Mounds must have been an important part of that visual language, and they needed to exercise a symbolism that would have been understood by both indigenous groups and incomers.

In the homelands during the Norwegian Iron Age, burial mounds were raised as territorial and political markers, as well as memorials of the important dead.[36] Mounds proclaimed right to land and signified a family of ancestral worth. Burial mounds continued to be important in the Viking period in Norway, with major mound cemeteries at Borre and Kaupang. Recently published work on the early proto-urban site of Kaupang demonstrates how mounds were an integrating and powerful element in an important settlement area: the surrounding landscape included the burial mounds of the cemeteries, a thing site, and a major hall mound.[37] At Borre, near the Oslofjord, mounded gravesites began to be erected in the seventh century; that their potency continued is obvious, as earlier mounds were despoiled and thus appropriated in the Viking period, perhaps as a dynasty fell out of favour. Some important Viking Age Norwegian buildings were also erected on mounds and prominent points, such as the hall at Skiringssal and the huge hall at Borg in Lofoten.[38] In both these cases, considerable effort was expended in rebuilding the hall structures in exactly the same spot on the most prominent part of the hillocks they occupied.

Layers in mounds have also been interpreted as part of ceremonies to bring together local populations in the process of creating new power structures. Terje Gansum and Terje Oestigaard interpret new excavation evidence on the huge mounds at Tønsberg that seems to demonstrate that the mounds were cenotaphs.[39] Although excavations have found no significant burials or grave goods despite being identified by historian Snorri Sturluson as the burial mounds of the sons of the first "king of Norway" Harald Finehair (c. 870-930), the layers of the mounds were examined in detail in the new excavations, with fascinating results. The mounds had been created in part by spreading and piling up enormous quantities of charcoal. But this charcoal had been brought in from elsewhere, and its generation and import must have demanded a great deal of work. Similarly, the soil and turf also used in constructing the mound came from different areas of the surrounding region, signalling the involvement of a considerable number of people as well as the drawing together of different groups. Thus, these mounds were associated with the power to control a number of groups of people and their landscape being exerted by a new regime. Even without burials, the mounds could be used to send potent signals. The mounds' continuing significance is illustrated by the later ascription of their origins to the resting place of a king's sons.

The symbolism of mounds was further reinforced by the holding of things (assemblies) at or on mound sites. It is hardly surprising that more mobile Viking groups in England sometimes referenced the significance of mounds by reusing prehistoric burial mounds, stamping their authority on the existing social landscape.

Mounds were equally potent in the settlement landscape of Orkney into which the Vikings moved. The coastal landscape so rich in economic potential was characterized by mounds: Neolithic tombs and coastal settlement mounds, Bronze Age enigmatic mounds of burnt stones and burial mounds, and Iron Age broch and burials mounds. Every Viking-Norse coastal settlement of significance is surrounded by and constructed on these past mounds. At Skaill, Deerness, Viking-Norse structures share the bay with a broch, settlement mounds, and middens, and with Bronze Age burnt mounds;[40] at Skaill, Sandwick, there were several prehistoric burial mounds in the centre of the bay, at least two visible broch mounds, and the coastal mounds of Skara Brae; at Westness (Skaill), Rousay, the landscape is littered with mounds and reminders of past settlement. Taking possession of a significant mound in a prime location, building a hall (Old Norse *skáli*) there, and further enhancing the visually dominant mound was to take possession of the landscape, to lay claim to the past of the area. An immediate link was created, with the existing sites of importance perpetuating a symbolism pertinent to new arrivals and indigenous populations alike. A longhouse on a mound monumentalized landscape control and economic power *and* the intention to add social ascendancy to physical dominance.

Central Place Theory: The Settlement Pattern

There are surprisingly few extensively excavated Viking-Norse settlements in the Northern Isles and perhaps as a result very few attempts to explore the social organization of the overwhelmingly coastal settlement pattern. Yet, settlement mounds are obvious candidates as focal or central places. They are, almost without exception, in the better and sandy locations on coasts and bays, ideally located to pursue the broad-based economy described above, and to supplement that with trading and raiding. Like Iron Age hill forts or brochs of the preceding period, the mounds created a dominating physical focus in the landscape. Most were placed to be visible around the sweep of their bays; others are sited principally to dominate the sea approach, like the Marwick mound.[41] In guiding craft to land they may also have proclaimed a right to the resources of the surrounding sea and coast. Thus, settlement mounds provided the central places around which the local social landscape was organized: local communities could navigate by them socially as well as geographically.[42] As discussed above, such a local communal focus would have been economically necessary. The character of the agricultural and grazing land certainly dictated scattered settlement, with each farmstead needing a spread of different types of land – grazing, hay, peat/turf, fields, garden area, and access to the shore and foreshore. But other environmental constraints impelled communities to collaborate: to fish, to build, to trade, and to maintain their herds. Local decisions about trading, fishing, or even participation in raids may have been made in the commodious hall on the mound, as well as resolutions about marriages and other local matters too small to take to a regional thing. The halls may have replaced local thing sites. It is interesting to note that in the well-settled archipelago of Orkney there are only two known thing site plane names, Dingieshowe and Tingwall; this may be an accident of survival but suggests perhaps that affairs often dealt with at such sites in other areas of the North Atlantic were, in Orkney, being settled at other locations.

Local communities were also required to offer up tribute-in-kind to support the earldom system, as well as to contribute to the formalized hospitality demanded by the earls and locally powerful men as they criss-crossed the islands.[43] Significantly, the Orkneyinga saga often refers to a group called "the farmers," which needed to be kept reasonably happy – presumably as the source of food, drink, and tradable goods – but is also found protesting at the burdens of tribute laid upon it, as one earl or another pursues campaigns of raiding or conquest.[44] In this context, the locally dominant mound settlement again acted as a central place location: collecting the crops, foodstuffs, and other goods needed for feasting, trade, or tribute.

Place-name evidence helps underpin this settlement model. Many of the settlement mounds, both excavated and simply recorded, are linked to Skaill places names (*skáli* in Old Norse). Elsewhere in the Viking-Norse North Atlantic, and particularly in England, *skáli* place names have been interpreted as meaning a hut or simple shelter. But in Orkney and in Norway, the place name *skáli* clearly refers to a hall and even implies it is one of some status. Indeed, the *skáli* place name seems to be a designation of function and role, rather than simply a descriptive name, or a name linked to a particular person or directly to a topographical feature.[45] *Skáli* is most often combined with *Lang* to make "Langskaill" – an important hall of some size. This meaning, implying a focal and important building, seems to have been used most consistently in Orkney, suggesting there was a clear need in the language of settlement for a place name capturing the idea of a central place-type location. In the Orkneyinga saga, the *skális* are connected to powerful people: earls and men like Thorkel Fostri and Sweyn Asleifsson.[46] The saga often uses the word *veizla* when describing the context for events held at these important *skáli*; the word *veizla* encapsulates the whole structure of giving, honouring, and personal loyalty linked to powerful men. So the *skális* were the stages for formalized hospitality, for *veizla* – for feasts given to seal alliances, affect political reconciliation, or heal breaches of trust; for the offering of gifts or tribute, and the establishing of bonds of honour. These were the events that would have required the local community to fulfill the demands of a formal hospitality. On such occasions, *skális* were the nodal points for the fulfillment of obligations to earls, bishops, and their powerful representatives. Skaill-type places were therefore the meeting point of local organization with the overarching earldom political structure.

Place-name evidence is often problematic to deal with: names move in an area, disappear, and lose their meaning. It can be no coincidence, however, that the coastal mounded landscapes so environmentally attractive to the Viking-Norse settlers are also disproportionately rich in Skaill and Langskaill names. Marwick Bay is a classic example. Two Skaill names crop up in the centre of a bay peppered with mounds of uncertain date and origin; on the coast is an extensive settlement mound, the upper layers of which have been dated to the Viking-Norse period, eroding into the sea and very close to an early chapel site.[47] At the Bay of Skaill, the excavated mounds, with high-status Viking settlement, are found near an early church site, St. Peter's Kirk, as well as the find-spot of Scotland's largest Viking silver hoard. I am now investigating all the Orkney Skaill sites. Of forty-two sites, the overwhelming number – there are only a couple of exceptions – are in these prime coastal mound-rich locations. The combination of mounded coastal landscapes, the vast majority of which have windblown sand-drift geology, with high-status Viking-Norse settlements on mounds and Skaill names, is extremely suggestive.

Conclusion

Economy, social organization, and landscape come together to demonstrate how, rather than being in the grip of hostile and limiting climes, the Viking-Norse inhabitants of northern environments exploited and manipulated their surroundings to reflect their society. The settlement landscape illustrates their technological appropriation of the surroundings to create a cultural environment. In Viking-Norse Orkney, social consequence was conveyed through the application of specific building technologies supported by sophisticated environmental know-how. The cultural and political landscape was given physical expression by referencing and reworking long-standing, potent symbolism in a way relevant to the Orcadian context. This settlement landscape is not evidence for dour survival but for a way of living that strengthened social organization by making creative use of essential landscape resources, as well as more obvious status symbols. Coastal mounds were monuments to economic and social success, acting as central place locations that both constructed local society and represented the local community. Creating and extending these mounds, and exploiting their powerful symbolism, relied crucially on a deep understanding of the full potential of a very particular northern environment.

NOTES

1 Anthony Giddens, *The Constitution of Society: Outline of the Theory of Structuration* (Cambridge, UK: Polity Press, 1984); Tim Ingold, *The Perception of the Environment: Essays in Livelihood, Dwelling and Skill* (London: Routledge, 2000).

2 Jane Harrison, *Working with Sand: Coastal Living in Viking and Late Norse Orkney* (MSc thesis: University of Oxford, 2007); Raymond Lamb, *The Archaeological Sites and Monuments of Sanday and North Ronaldsay*, RCAHMS List no. 11 (Edinburgh: Royal Commission on the Ancient and Historical Monuments of Scotland, 1980).

3 Ian Armit, ed., *Beyond the Brochs* (Edinburgh: Edinburgh University Press, 1990); Anna Ritchie, "Excavations of Pictish and Viking-Age Farmsteads at Buckquoy, Orkney," *Proceedings of the Society of Antiquaries of Scotland* 108 (1977): 174-227.

4 John R.C. Hamilton, *Excavations at Jarlshof, Shetland* (Edinburgh: HMSO, 1956).

5 Birsay sites: John W. Hedges, "Trial Excavations on Pictish and Viking Settlements at Saevar Howe, Birsay, Orkney," *Glasgow Archaeological Journal* 10 (1983): 73-124; Christopher D. Morris, *The Birsay Bay Project: Coastal Sites beside the Brough Road, Birsay, Orkney: Excavations, 1976-1982*, vol. 1 (Durham, UK: University of Durham, Department of Archaeology, Monograph Series, 1989); Christopher D. Morris, *The Birsay Bay Project: Sites in Birsay Village and on the Brough of Birsay, Orkney: Excavations, 1976-1982*, vol. 2, Monograph Series 2 (Durham, UK: University of Durham, Department of Archaeology, 1996).

Skaill, Deerness: Simon Buteux, *Settlement at Skaill, Deerness, Orkney*, British Series 260 (Oxford: British Archaeological Report, 1997).

Bay of Skaill: Data Structure Reports (Archaeological Reports), Birsay-Skaill Landscape Archaeology Project (B-S LAP). Unpublished data structure reports: Annual University of Oxford archaeological reports for the B-S LAP, 2004-1.

Westness: Sigrid H.H. Kaland, "Some Economic Aspects of the Orkneys in the Viking Period," *Norwegian Archaeological Review* 15 (1982): 85-95.

Tuquoy: Olwyn A. Owen, "Tuquoy, Westray, Orkney," in *The Viking Age in Caithness, Orkney and the North Atlantic*, ed. Colleen Batey, Judith Jesch, and Christopher D. Morris (Edinburgh: Edinburgh University Press, 1993), 318-39; Olwyn Owen, ed., *The World of Orkneyinga Saga: "The Broad-Cloth Viking Trip"* (Kirkwall, UK: Orcadian, 2005).

Pool: John R. Hunter, Julie M. Bond, and Andrea N. Smith, eds., *Investigations in Sanday, Orkney*, vol. 1, *Excavations at Pool Sanday, a Multi-Period Settlement from Neolithic to Late Norse Times* (Kirkwall, UK: Orcadian/Historic Scotland, 2007).

Shetland sites: Stephen J. Dockrill et al., *Excavations at Old Scatness, Shetland*, vol. 1, *The Pictish and Viking Settlement* (Lerwick, Scotland: Shetland Heritage Publications, 2010); Julie Bond, "Ashes to the Earth: The Making of a Settlement Mound," in *Old Scatness Broch, Shetland: Retrospect and Prospect*, ed. Rebecca A. Nicholson and Stephen J. Dockrill (West Yorkshire, UK: North Atlantic Biocultural Organisation, 1998), 81-96.

6 On Ireland and the Irish Sea zone, see David Griffiths, *The Viking of the Irish Sea: Conflict and Assimilation, AD 790-1050* (Stroud: History Press, 2010).

7 Hunter, Bond, and Smith, *Investigations in Sanday, Orkney*; Ritchie, "Excavations of Pictish and Viking-Age Farmsteads"; and Dockrill et al., *Excavations at Old Scatness*.

8 Reidar Bertelsen, "Farm Mounds in North Norway: A Review of Recent Research," *Norwegian Archaeological Review* 12 (1979): 48-56; Reidar Bertelsen and Raymond G. Lamb, "Settlement Mounds in the North Atlantic," in Batey, Jesch, and Morris, *The Viking Age in Caithness, Orkney and the North Atlantic*, 544-54; and Lamb, *Archaeological Sites and Monuments.*

9 Reidar Bertelsen, "Farm Mounds of the Harstad Area," *Acta Borealia* 1 (1984): 7-25.

10 William W. Fitzhugh and Elizabeth I. Ward, eds., *Vikings: The North Atlantic Saga* (Washington, DC: Smithsonian University Press, 2000); and Dagfinn Skre, ed., *Kaupang in Skiringssal.* Kaupang Excavation Project Publication Series, vol. 1 (Aarhus, Denmark: Aarhus University Press, 2007).

11 Morris, *The Birsay Bay Project: Sites in Birsay Village*; Data Structure Reports (DSRs), Birsay-Skaill Landscape Archaeology Project.

12 Morris, *The Birsay Bay Project: Coastal Sites.*

13 On this specialization, see Christopher E. Lowe, *Coastal Erosion: The Archaeological Assessment of an Eroding Shoreline at St Boniface Church, Papa Westray, Orkney* (Stroud, Scotland: Historic Scotland and Sutton Publishing, 1998).

14 Alan Davidson, *North Atlantic Seafood* (London: Penguin Books, 1980).

15 DSRs, Birsay-Skaill Landscape Archaeology Project.

16 Hunter, Bond, and Smith, *Investigations in Sanday, Orkney.*

17 David Griffiths and Jane Harrison, with Michael Athansan, *Birsay-Skaill Landscape Archaeology Project, Orkney, Phase XI: Data Structure Report 2009; Marwick Bay MW09*, University of Oxford report, *DSR 2009* (Oxford, 2009).

18 James H. Barrett, "Culture Contact in Viking Age Scotland," in *Contact, Continuity and Collapse: The Norse Colonization of the North Atlantic*, ed. James H. Barrett (Turnhout, Belgium: Brepols, 2003), 73-111.

19 Hermann Palsson and Paul Edwards, *Orkneyinga Saga* (London: Penguin, 1978); Owen, *World of Orkneyinga Saga.*

20 DSRs, Birsay-Skaill Landscape Archaeology Project.

21 David Griffiths and Jane Harrison, *Birsay-Skaill Landscape Archaeology Project, Orkney: Data Structure Report 2004-5, BS04 (Birsay) and SG04 (Snusgar)*; *DSR 2004*, University of Oxford report (Oxford, 2005); David Griffiths and Jane Harrison, *Birsay-Skaill Landscape Archaeology Project, Orkney, Phase IV: Data Structure Report 2005, SG05 (Snusgar); DSR 2005*, University of Oxford report (Oxford, 2006); David Griffiths and Jane Harrison, *Birsay-Skaill Landscape Archaeology Project, Orkney, Phases V, VI and VII: Data Structure Report 2006, SG06 (Snusgar), Survey and Excavation Phase VI; DSR 2006*, University of Oxford report (Oxford, 2007); and David Griffiths and Jane Harrison, *Birsay-Skaill Landscape Archaeology Project, Orkney, Phases VIII, and IX: Data Structure Report 2007, SG07 (Snusgar: DSR 2007)*, University of Oxford report (Oxford, 2008).

22 David Griffiths and Jane Harrison, *Birsay-Skaill Landscape Archaeology Project, Orkney, Phase XII: Data Structure Report 2010, SG10 (Snusgar): DSR 2010*, University of Oxford report (Oxford, 2010).

23 *DSRs 2007-8; DSR 2010;* and David Griffiths and Jane Harrison, *Birsay-Skaill Landscape Archaeology Project, Orkney, Phase XIII: Data Structure Report 2011, SG11 (Snusgar); DSR 2011*, University of Oxford report (Oxford, 2012).

24 On the smithing see *DSR 2010.*
25 *DSR 2009;* Harrison, *Working with Sand.*
26 J.D. Hill, *Ritual and Rubbish in the Iron Age of Wessex: A Study on the Formation of a Specific Archaeological Record* (Oxford: Tempus Reparatum, 2005).
27 David McOmish, "East Chisenbury: Ritual and Rubbish at the British Bronze Age-Iron Age Transition," *Antiquity* 70 (1996): 68-76; and Stuart Needham and Tony Spence, "Refuse and the Formation of Middens," *Antiquity* 71 (1997): 77-90.
28 Hunter, Bond, and Smith, *Investigations in Sanday, Orkney.*
29 *DSR 2008.*
30 *DSR 2012.*
31 Iain Crawford, "Archaeological Prospect and Practical Technique in an Environmental Region: The Western Isles Machairs," *World Archaeology* 10 (1978): 51-62.
32 Harrison, *Working with Sand.*
33 Reidar Bertelsen has investigated how building on agricultural midden may have helped raise the internal temperature of Norse turf houses in northern Norway in Reinhardt Mook and Reidar Bertelsen, "The Possible Advantage of Living in Turf Houses on Settlement Mounds," *Acta Borealia* 24 (2007): 84-97.
34 *DSR 2010.*
35 Ritchie, "Excavations of Pictish and Viking-Age Farmsteads."
36 Fitzhugh and Ward, *Vikings.*
37 Skre, *Kaupang in Skiringssal.*
38 Gerd S. Munch, Olav S. Johansen, and Else Roesdahl, *Borg in Lofoten: A Chieftain's Farm in North Norway* (Trondheim: Tapir Academic Press, 2003).
39 Terje Gansum and Terje Oestigaard, "The Ritual Stratigraphy of Monuments That Matter," *European Journal of Archaeology* 7 (2004): 61-79.
40 Buteux, *Settlement at Skaill, Deerness, Orkney.*
41 *DSR 2009.*
42 Charlotte Fabech and Jytte Ringtved, eds., *Settlement and Landscape: Proceedings of a Conference in Arhus, Denmark, May 4-7, 1998, Moesgård* (Denmark: Jutland Archaeological Society, 1999).
43 William P.L. Thomson, *The New History of Orkney*, 3rd ed. (Edinburgh: Birlinn, 2008); and William P.L. Thomson, *Orkney: Land and People* (Kirkwall, UK: Orcadian Ltd./Kirkwall Press, 2008).
44 Palsson and Edwards, *Orkneyinga Saga.*
45 Raymond G. Lamb, "Historical Background to the Norse Settlement," in Buteux, *Settlement at Skaill, Deerness, Orkney*, 13-16; Thomson, *Orkney: Land and People.*
46 For example, Palsson and Edwards, *Orkneyinga Saga*, chap. 16, dealing with Earl Einar and Skaill, Deerness; and chaps. 65 and 74 dealing with Earl Paul Håkonsson and Skaill, Rousay.
47 *DSR 2009.*

In Search of Instructive Models
The Russian State at a Crossroads to Conquering the North

JULIA LAJUS

Mastering the North was a long-term problem for the Russian state, which at least from the eighteenth century tried to organize the effective use of its resources. Even the definition of the North, as the region with limited agriculture and extensive aquatic resources and forests, was resource based. Colonization from the very beginning was led by the search for these resources: forest hunting, salmon fisheries in the rivers, marine hunting, and marine fisheries supplemented each other, with one or two activities being more typical in each area.

In this chapter I illustrate two very distinct foreign models employed for the "state colonization" of the Russian North in a formative period between the Great Reform of 1861 and Stalin's industrialization of the 1930s: Norway and Canada. Although the use of the Norwegian model for colonization of the Russian North is relatively well studied, "railway colonization" of the 1920s is not that well known, and very few works embrace both imperial and early Soviet periods of colonization.[1]

The search for foreign models, technologies, and institutions that might be adopted for the development of the colonization process was a characteristic feature of Russian colonial governance.[2] Although the colonization of that period is well studied for the eastern and southern Russian borderlands, the North does not usually get its well-deserved place in these discussions, especially because it gives us very little material for studying inter-ethnic conflicts. However, the evolving and multi-faceted process "in which outsider colonists, native peoples, the natural environment, and the word of the state and its representatives influenced one another in ever shifting combinations" was equally pronounced in the North.[3]

The colonization process in the Russian North, especially in its most western part, known as the Kola North – situated at the Kola Peninsula and washed by the White Sea in the south and east and by the Barents Sea in the north – in comparison with other Russian periphery was complicated by its physical and mental closeness to the enlightened Western world (Map 5.1). This latter circumstance provided an additional temptation to borrow foreign models. However, in attempts to implement models, as usually result from contacts with the counterpart, with the Other, Russians first and foremost gained knowledge about themselves, facilitating better understanding of the political and geographical differences of the Russian North from the North of other northern countries.[4]

Highly centralized Russian and Soviet empires made numerous attempts to colonize the North and technologically and socially modernize its traditional resource-based economy. For instance, under Peter the Great, colonial-type policy was developed with the application of the Dutch model of large trade monopolies.[5] Although Peter the Great dreamed about improving the use of marine resources in the northern seas, he sacrificed the international trade route through Arkhangelsk for the rise of St. Petersburg. This decision weakened the North, which became a periphery in position not only to Russian capitals but also to Europe.

By the end of the nineteenth century, the place of the Russian North on the mental maps of the authorities and intellectuals of the Russian Empire was very distinct from the place of the North in Scandinavian countries and Canada. In contrast to the Nordic notion of the North, which "became a dimension of the formation of a modern Swedish self-understanding, and indeed nationalism,"[6] or the Canadian "almost unbounded optimism" that guided their northern approaches in the second half of the nineteenth century,[7] the Russian North lost its cultural and economic significance with the southeastern expansion of empire. It was considered as a periphery without clear perspectives of economic development. The significance of the Russian North was gradually and partly recognized only with the rapid rise of the Scandinavian North at the very end of the century.

There is no clear answer to the question of whether Russia had a northern policy at the end of the nineteenth century. It is evident that the policy was weak, changeable, and generally defensive.[8] The North usually became an issue only in the face of real or hypothetical threats from other countries,

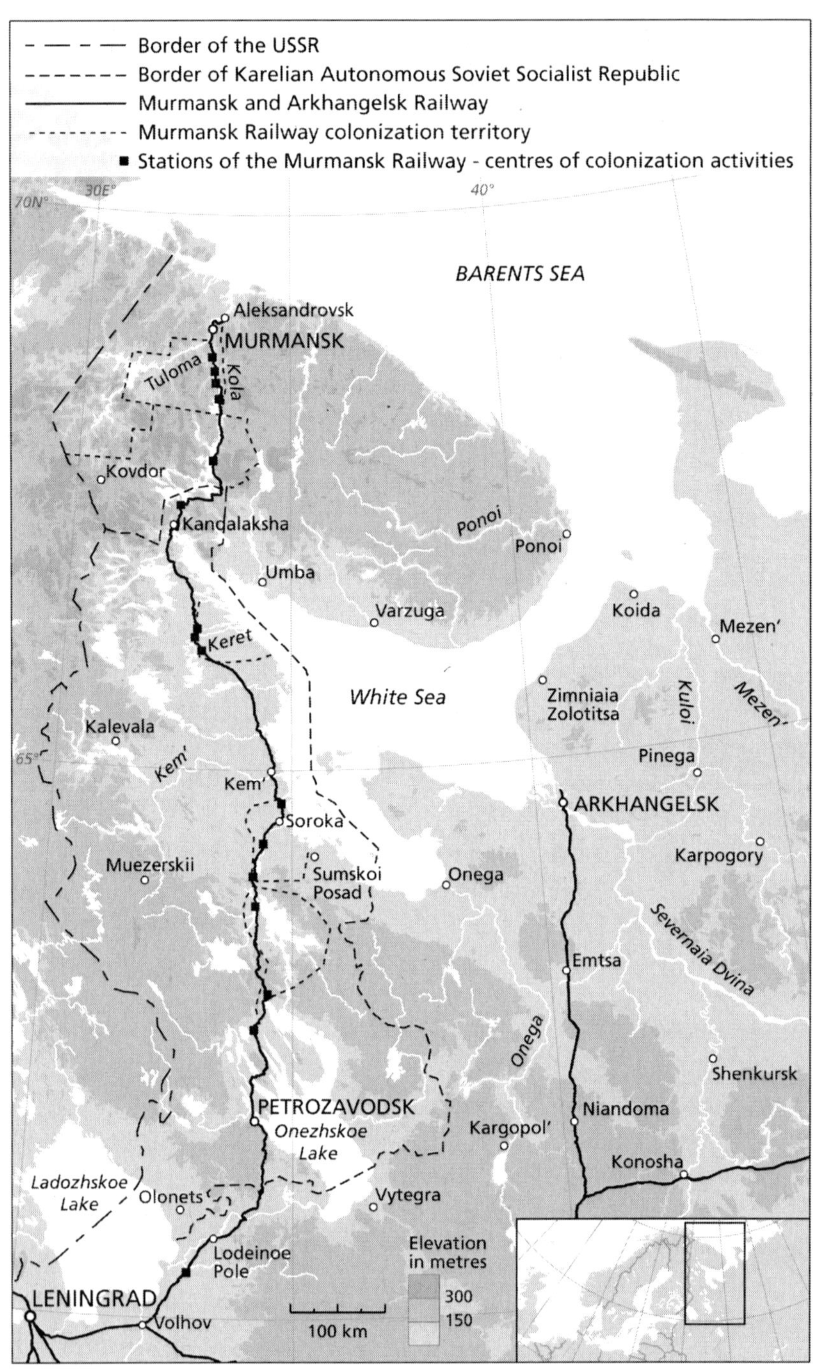

Map 5.1 The Russian Kola North

with fear of territorial loss. Also it seems that at least since the mid-nineteenth century, with the total decline of Russian marine hunting activities both at Novaya Zemlya/Spitsbergen and in the North Pacific and by the Stalinists' turn to the Arctic, the state did not create and support a pronounced northern identity for Russia. There were some splashes of interest toward the North connected with the Spitsbergen question, but they seem to have been relatively weak.[9] Weakness of both northern identity and northern policy, together with the understanding of the backwardness of Russian northern periphery in comparison with neighbour countries, forced the state to refer to foreign examples in its efforts to develop the region.

Around the turn of the nineteenth century, the instructive examples were sought mostly in Scandinavia, and especially in Norway. But in the early Soviet times of the 1920s, the future of the North, which was being rapidly reshaped with technology such as railroad construction, was described as becoming a "Soviet Canada." The peculiar term "canadization" was coined to name the new methodology of colonization of remote northern territories. It referred to the multi-sided development that came along with the North American railways.

The search for instructive models came to an end with militarization of the region, especially the organization of the naval Northern Fleet in 1933, which got under its authority the best fisheries bays on the Murman coast. Simultaneously, the emergence of the Arctic Empire of the Main Administration of the Northern Sea Route facilitated further expansion of the newly formed Soviet style of dealing with space, environment, and resources expanded into the North. Rapid movement into the High Arctic brought the sub-Arctic North under the Soviet policy of internal colonialism with the development of large-scale industrial sites and gulags.[10]

Looking to Norway: A Model from the Nearest Neighbour

The colonial attitudes toward the North, which were characteristic of Peter the Great's policy by the end of the eighteenth century, changed to understandings of the North as a periphery that could operate according to the general rules of the country and did not need special attention and care from the state. Thus, from the late eighteenth century to the mid-nineteenth, state intervention in the economic activity of the northern inhabitants, who lived along the coasts of the White Sea and were known as Pomors, was rather insignificant. However, this situation could not continue for long in the framework of imperial development of the country. The very unsuccessful Crimean War revealed the particular weakness of the Russian North, where the coast of the Barents Sea, with its almost complete lack of permanent settlements, could be easily accessed by the European powers. In 1859, the Ministry of State Domains sent a scientific expedition under the leadership of Nikolai Danilevsky to explore possibilities for the economic development of the Russian North, including colonization of the Murman coast of the Barents Sea.[11]

Rapid development of northern Norway, especially the province of Finnmark, which had become prominent since 1840, strongly contrasted with the lack of development on the Russian side of the border. In Finnmark, a coastliner service had been established in 1838, fisheries were blooming, and the ice-free conditions facilitated international trade connecting Finnmark to southern markets as far away as Italy and even Brazil.[12] By 1870, the telegraph extended its service to all the major settlements in the province. All these measures significantly facilitated immigration, which increased the population of Finnmark fourfold over the course of the nineteenth century. Because of these achievements, "following the Norwegian example" and "reaching the Norwegian level" became commonplace phrases in discussions on the future of the Russian European North.

Russian officials and travellers who visited northern Norway were struck by the contrast between it and the Russian North, which could not be explained by environmental factors alone, as the border between Norway and Russia did not follow any environmental barrier.[13] Although Russians complained that they lost some good fishing grounds to Norway, there still were the ice-free bays, excellent for fisheries, on the Russian side. The Murman coast was a natural extension of northern Norway, where commercial fish such as cod, halibut, and herring migrated with the warm Gulf Stream waters. These migrations, however, were variable from year to year, thus the catches were not stable and differed between western and eastern Murman coasts. Not going into these important details, the initiator of governmental colonization of the Murman coast, the Russian Consul General Christiania Leo Mekhelin, overemphasized the environmental similarity between Finnmark and this part of Russia: "Russian Lapland, being situated adjacent to its Norwegian portion, is similar with this latter in terms of climate, population and livelihood."[14]

This statement, not even fully true regarding the climate, was definitely wrong regarding the population and livelihood. One of the main differences with Scandinavia and a main obstacle for Russian colonization was the traditional use of the Murman coast for seasonal fisheries in which only men, who came there in spring from the villages on the White Sea coasts, five to six hundred kilometres away, participated.[15] Women in Pomor culture were very much connected with "home"; for them it was crucial to maintain close relations with the local community, relatives, and church. Because of this, the colonization process was contained to areas where relatively large villages could survive.

The Pomor economy was the only one in traditional Russia based on the use of marine resources, and even in this economy the rivers and river fisheries played an important role. Seasonal cod fisheries at the Barents Sea coast developed rather late – not earlier than the beginning of the sixteenth century, and already at that time, most of them probably developed on the basis of technological transfer from the Dutch and British fisheries cultures. Thus, at the end of the nineteenth century, the Pomor use of marine resources stood far from the core of Russian patterns of resource use based exclusively on land, forest, and river. That it was no longer common to use the resources of the open sea, together with the economic organization of Pomor life as a peasant commune, hindered the colonization of the Barents Sea coast, where environmental conditions demanded another pattern of colonization – by single families or very small communities – typical of Norwegian and Finnish colonists.

The outlier features of the Murman coast and its people placed the north of Russia regions in a particular place on the mental maps of authorities and intellectuals. Whereas for Norway the development of Finnmark was an important task related to the strengthening of its border with Russia and the "Norwegization" of northern minorities, as well as crucial to improving fisheries and increasing trade, for the enormous Russian Empire the Murman coast was a "bear corner" without clear economic perspectives and very limited strategic interests.

The Murman coast's proximity to small Norway, a part of Sweden at that time, was seen as an advantage, because it would provide a model for improvement. Not only that; it could also provide actors who would bring this model directly to Russia. The governor of Arkhangelsk province, who supported the idea of colonization of the Murman coast, emphasized this possibility: "It would be more expedient to offer the opportunity of settling on the Murman coast not to Russians alone, but also to Norwegians, as this would serve toward encouraging the Russian fishermen in their work also, who, seeing no model, may currently not make the decision to settle on a deserted coastline."[16] Thus, inviting Norwegians and Finns to settle on the coast, the Russian authorities expressed the utopian idea that foreign colonists could serve as an encouraging example for reinforcing Russian colonization. Looking to the neighbour was interlinked with looking to the past, especially in such narratives where Russian northern fishermen of the eighteenth century were described as being very industrious and having extended their fisheries even into Norwegian waters.[17] It was thought that Norwegians would not bring something new but rather revive the glorious old days of Russian northern fisheries.

Opposition came from the naturalist Nikolai Danilevsky, who knew the region through expedition work and who had also spent several months in Norway studying its fisheries. To his well-trained gaze, the environmental similarities looked too obvious in general and not fully true in details, especially for the eastern Murman, where the climatic conditions were harsher and the bays not so convenient for fisheries as in northern Norway. However, his main concerns came from another direction, as he did not believe that the Norwegian instructive model would work. In general, he argued that there were fundamental differences between Russian and Western "cultural-historical types," an opinion he expressed later in the 1869 seminal philosophical book *Russia and Europe*.[18] Danilevsky doubted the usefulness of any Western models for the development of Russia. He did, however, agree with the proposal to allow specified numbers of Norwegians to settle on the coast. In that case, he argued, the permanent settlements of the Pomors might not be needed; it would be enough if the Pomors engaged in business interactions with the Norwegians.[19] He also encouraged the use of Norwegian types of fishing boats, and on the basis of his recommendations, the Ministry of State Domains purchased several of these boats and distributed them to the most experienced Russian fishermen.

The reality did not meet these expectations: Norwegians preferred to live separately from Russians, keeping their own cultural traditions; most of them did not learn Russian; and the main business in which they interacted was trade, including selling alcohol, which was the most profitable business for them.

When the first illusions dissolved, the central and regional Russian authorities changed their attitudes toward the presence of Norwegians on the coast and simultaneously toward the Norwegian model of economic development. They began to pay ever greater attention to the real and hypothetical harm caused by foreigners who settled on the coast. By the 1880s, it became increasingly clear that improvement was hampered by the very slow development of infrastructure; the lack of a juridical base for developing credit and insurance systems to stimulate private business in fisheries; and low attractiveness of the remote Barents Sea coast for the settlers, in spite of some government subsidies, tax, and recruit privileges.

A project for developing a naval base at Kola Bay, which would have been connected by railway with the Russian capital, was proposed in the 1890s by the influential minister of finance, Sergei Witte, but rejected as being too costly. Naval circles that might have been interested in the development of the only free entrance to the ocean for European Russia were not familiar with the North. The core of the nineteenth-century navy elite came from the Baltic German nobility, and naval policy in general was directed to the development of the Baltic and Black Sea ports.

Failure to build a railway isolated the Kola North – it could not become a part of a viable "imperial space," which had undergone rapid modernizing colonization. In other parts of Russia, even northern ones like Arkhangelsk and Vologda, dwellers could petition the authorities for a railroad.[20] Because the definition of "the North" was so broad and covered a very large territory, even citizens of Vologda were able to talk about "the far North" in their spatial rhetoric when they lobbied for the railway construction, arguing that their town was located in a key position to connect central Russia with the "Northern region."[21] Absence of a significant town in the Kola North left the discussion of possible railroad construction there to the central authorities and public in the centres only. Difficulties in reconceptualizing the Kola North as an integral part of Russia warmed up the desire to look toward Norwegian models for its development.

In spite of all the problems, several measures to improve the infrastructure along the Murman coast following the Norwegian example were undertaken in the last two decades of the nineteenth century: the ship route with Arkhangelsk was improved, making passenger transportation on a regular basis possible; a new administrative centre and port of Aleksandrovsk was opened in 1899; and the telegraph connected all major fisheries stations.[22] Moreover, one crucial component of the Norwegian model to which Russians referred most often in the 1890s was added to the picture: application of scientific knowledge.

For this knowledge Russian authorities and scientists proposed to look also to Scandinavia, which was a leader in modernization of marine resource harvesting. For instance, Scandinavian scientists led the establishment of the International Council for the Exploration of the Sea in 1902.[23]

Not only was science itself blooming in Scandinavia but so was its practical application.[24] Using science to solve practical problems in medicine, agriculture, and industry was becoming usual practice in Russia as well; it was discussed both at the universities and at the sessions of numerous societies where experts and practitioners met, for example, at the meetings of the Russian Society of Pisciculture and Fisheries, founded in 1881. There, as well as within the Society for the Assistance to the Trade Navigation, also based in the capital, possibilities for the development of Murman fisheries in connection with colonization policy and the use of science were vividly discussed.[25] In addition to this utilitarian side of science, scientific exploration as a tool of appropriation of remote northern places had a long tradition in Russia. In that respect, it was quite similar to the patterns of the use of science in accord with the statecraft in Sweden and also in Canada.[26]

Members of these groups argued, pointing again to the example of Norway, that the fishermen needed knowledge of where, how, and when the fish came, and such knowledge was available only from scientists.[27] In such narratives, Russian fishermen were depicted as an object for education. They were considered to be too backward to improve their fisheries themselves, yet sufficiently adaptable so as to accept and follow instructions provided by scientists, who in turn needed to get these instructions from abroad. And a new generation of scientists was very keen and well equipped to do just that. Zoologist Nikolai Knipowitsch, who was asked to lead the Murman scientific fishery expedition – the first large scientific enterprise in the region, funded by the philanthropic Committee for the Aid to Pomors and organized in St. Petersburg – relied heavily on the emerging ideas and methods of international, and above all Scandinavian, fishery science and oceanography. His expedition in turn became the major Russian initiative to participate in oceanographic research under the umbrella of the International Council for the Exploration of the Sea.[28]

Support by Russian authorities for the development of fisheries scientific research was largely based on references to scientific involvement in solving fisheries problems in foreign countries such as Norway and Denmark.[29] Russian proponents of developing research emphasized that Russia was behind other countries in its level of scientific knowledge. As was so common in Russia during this period, the practical applications of research were not clearly formulated, the results considered as self-evidently useful.[30] There was no consensus on what type of fisheries to develop: while the scientists were asked to support small-scale inshore artisanal fisheries, they dreamed about developing large-scale offshore industrial fisheries. Thus, although the expedition attempted to introduce new types of fishing boats, which were mostly Norwegian types, to the fisherman, it also explored offshore fishing grounds in its research vessel equipped with a trawling net.

The North was the first region in Russia where local fishermen directly encountered new industrial fisheries. In the 1900s, British trawlers appeared near the Murman coast, attracted by reported positive results of experimental trawling by Murman expeditions. Soon, two Russian industrialists bought trawlers as well.[31] The future development of fisheries was actively discussed at meetings of the local Arkhangelsk Society for the Study of the Russian North, established in 1908.[32] Industrialists and scientists influenced by the British example advocated for attracting capital investment for the development of industrial fisheries based on trawling. Another group of local representatives and some scientists supported small-scale inshore fisheries and argued against the trawl, as in the Norwegian model.

In the 1910s, the invasion of English trawlers, along with Norwegian fishing vessels and marine hunting expeditions, in a region where no formal borders of territorial waters had been established, redirected the attention of central and local authorities from their concerns about colonization of the coast and modernization of fisheries toward concerns about guarding the region from foreigners. It also became evident that scientific knowledge by itself could not help; significant changes in socio-economic organization and technology were needed also.

"Soviet Canada"

After 1917, the new Soviet power strongly emphasized the role of science and technology in building Communism. The effective use of railways was in line with this policy.[33] Thus, in 1923, with the development of the New Economic Policy, which declared relative liberalization of economics in comparison with the War Communism of the first years after the revolution, the Murmansk Railway was transformed into an economic body of a new type, called a "combinat" (industrial complex) – a "combination of several technologically connected specialized industrial plants of different branches of economy."[34] According to the governmental decree of 25 May 1923, the railway got more than three million hectares of land, the aim being to use it for the purposes of the colonization and economic development of the region.[35] The Soviet government defined colonization as "an involvement into economic circulation of necessary lands with the purpose of increasing agricultural and industrial production by the means of rational settlement policy and rational exploitation of natural resources."[36]

The term "canadization" was devised to name this new method of colonization of remote northern territories, referring to the experience of the North American railways in the multi-sided development of the regions they ran through. When these railway companies were founded, they got a large amount of land for developing agriculture, forestry, and industry to produce goods for transport and to attract labour. Soviet officials emphasized the critical role of railroads in North American settlement: "[In] contrast to Europe, where different nations built their railways, in America, the railway system created the American nation."[37]

The Soviet sources constantly referred to the Canadian Pacific Railway – the first transcontinental railway in Canada, built in the 1880s – which was granted twenty-five million acres of land by the federal government.[38] Based on these land resources, Canadian Pacific organized an intense campaign to bring immigrants to Canada and thus facilitated the colonization of the Canadian wilderness.[39] From the end of the 1890s, the Canadian Northern Railway, which branched off from the Pacific, stretched its new lines into remote northern territories, proving to be an important instrument for creating a new environment in the North.[40]

The Murmansk Railway was a perfect object for canadization. It had been built recently – during the First World War – with the strategic purpose of transporting military freight arriving from the Allies to the capital and then to the front. It ran through poorly populated regions rich in natural resources. As explained by the head of the Murmansk Railway's Colonization Department, Gennadii F. Chirkin, "In order for the railway to fulfill its purpose and overcome its essential challenges, it has to create its own cargo, thereby bringing to life this dormant region."[41] The perception of the North as a dormant region needing to be awakened by technology mirrored the perceptions in Canada, where the land was described as "being lonely" before the arrival of technologically equipped humans.

The most obvious resources of the "dormant region" were forest and fish. These resources are reflected in the names of the main industrial bodies of the new railway combinat: one was named Zhelles [Railway Timber], the other, Zhelryba [Railway Fish]. The new structure of the Murmansk Railway consisted of five "inseparable parts": the railway as a transport enterprise; Zhelles, the department that developed the forestry and timber industry; Zhelryba, the department that organized fisheries and, thus, in addition to earning money by the fish trade, provided a food supply to other departments; the Colonization Department, which recruited labour and organized the resettlement of people; and the Murmansk port.[42] The port began its

Figure 5.1 View of the town of Murmansk from Kola Bay, 1928. Visible are the final segments of the Murmansk Railway; timber at the shore of Kola Bay; the "hotel" of railway fisheries enterprise (Zhelryba) in the centre, behind the wooden railway building; and a large consumer's cooperative shop for transport workers behind the "hotel." Courtesy of Murmansk Museum of Regional Studies (photo number 15433/148-9).

industrial life when the railway reached ice-free Kola Bay at the end of 1916, which was the goal of the whole "titanic undertaking" of the railway's construction. After the end of the war, the importance of the port decreased, and it was the railway that was called on to re-energize the port (Figure 5.1). In addition, the Murmansk Railway became a major route for transporting Barents Sea fish to Petrograd/Leningrad, very similar to the Canadian Pacific Railway, which opened new markets for fish caught in British Columbia.[43]

Strikingly, the same people who developed pre-revolutionary colonization policy supported the new direction toward canadization. The major proponent of canadization was Gennadii F. Chirkin (1876-1938), in the 1920s the head of the Murmansk Railway's Colonization Department and the former head of the Resettlement Administration in the pre-revolutionary Ministry of Agriculture.[44] Historian Peter Holquist recently argued that "the technocratic, etatist agenda of this ministry was an important feature of state policy in the last years of the imperial regime and throughout the years of World War I."[45] He convincingly showed that the "technocratic ethos" was a crucial bridge for many specialists to serve the Soviet state, specifically referring to Chirkin to illustrate this thesis. Appealing to Western models of railway construction, including American ones, played a very important role in the discourse of engineers and officials who promoted these new technologies to Russian authorities in the mid-nineteenth century.[46] The Russian minister of communications, later the powerful minister of finance, Sergei Witte, referred to the experience of Canada, and the Canadian Pacific Railway in particular, when he argued for the construction of the Trans-Siberian Railway, especially its significance for the development of agriculture.[47]

In the period after the 1905 revolution when the Resettlement Administration was reorganized and moved under the authority of the new Main Administration of Land Management and Agriculture (GUZZ), its specialists worked in Siberia, far east Turkestan, and other regions. The north of the North, not being an agricultural region, did not receive attention before the completion of the Murmansk Railway radically changed the situation. Earlier Chirkin had written about the role of railroads in developing Siberia, so it was logical that he argued the same for the underdeveloped and underpopulated North.[48] He named colonization as the "alpha and omega of economic revival of the Russian North" and argued for the railway concession based on canadization, which should connect the Russian North with Siberia.[49]

Although the pre-revolutionary institution was named the Resettlement Administration, its specialists adopted from the repertoire of western European practice the term "colonization," in contrast to the more neutral term "resettlement," defining it as more state-directed and programmatic.[50] They systematically studied foreign experience in colonization, including the Canadian success. Some of them might even have heard a speech by Canadian high-level official and former minister of railways and canals (1879-84) Charles Tupper at the International Railway Congress in 1892, which took place in St. Petersburg.[51] After the revolution, the Resettlement Administration was renamed the Colonization Department of the People's

Commissariat of Agriculture (the successor of the Ministry of Agriculture, with two-thirds of its pre-revolutionary staff). During the 1920s, the term "colonization" was compatible with the Soviet course of development of the periphery of the country, whereas later the Soviet officials completely returned to the term "settlement," referring to colonization only for the practices of imperialism.[52]

In 1922, a new organization that was to play a leading role in promoting the "canadization project" was organized: the State Colonization Research Institute. In addition to the staff of this institute, the railway administration employed scientists from other institutes to do the research it considered most necessary for preparing the base for its colonization efforts, "with the purpose of mastering the non-living riches which were provided to the railroad, to define the possibilities and methods for their exploitation."[53] The research developed in several directions: experimental agriculture in sub-Arctic conditions, improvement of fisheries and fish products technology, and the search for mineral deposits.

The appropriation of land by the railway caused tensions with the local authorities, especially with the strong Karelian government. Its struggle resulted in changes to the initial plans, and the railway got rights only to timber areas that were not already in use and located in the centre and north of Karelia, rather than in the better-developed southern territories.[54] Skeptics used to joke about canadization, calling it the "Ryazansko-Canadsky experiment," prefacing the word "Canada" with the name of one of the most ordinary Russian provincial towns, Ryazan.[55] However, soon the whole region got the unofficial metaphorical name of "Soviet Canada." As Chirkin argued, "Although our first experiment in 'canadization' of the railway is being conducted in natural and environmental conditions that are very distinct from Canadian ones (the base of the Canadian economy is agriculture), it is, however, possible to name the Karelo-Murman region in connection with the Murmansk Railway's multi-sided work here as the Soviet Canada."[56]

Not only were differences in the natural and environmental conditions emphasized but also a crucial difference in the political regimes of the USSR and Canada (the latter, however, often resembled just a remark that was necessary to include for political reasons only). Thus, in 1928, Chirkin wrote:

> Now that the method of the colonizing work of the Murmansk Railway has had its practical introduction into life and its own face has been revealed for several years, it is possible to state that the analogy with the work of the

> Canadian railways does exist insofar as the scheme of colonizing work, but the difference in juridical and socio-economic conditions between Canada and our country is so large that it is possible to speak about the colonization work of the Murmansk Railway as constituting absolutely independent methods and practices of colonization, first applied here, though designed with consideration of the experience of Canadian railways.[57]

It is interesting to note that the whole discourse of these advertisements and reports on new Soviet colonization of the North in the 1920s was very distinct from what we know about Soviet propaganda of the conquering of the Arctic and mastering of nature in general.[58] In the 1920s, it was still possible to not refer to the revolutionary activities of the masses led by genuine Bolsheviks but to compare the Soviet colonizers with the heroes of Jack London: "To fulfill all this titanic work it was necessary to be people of a very different nature, to be the people we adore so much when we read stories about the pioneers and conquerors of Alaska, Klondike and Yukon!"[59] This comparison also became a component of the term "pioneering-colonizing railway project," which was widely in use. Such discourse served to include the Soviet North in the common cultural realm of circumpolar space, where strong people conquered the hostile environment by the means of human spiritual power, merged with a new technology.

The well-known journalist Vasily Lokot (1899-1937), who published his reports and satirical stories under the pseudonym A. Zorich, published *Sovetskaia Kanada* (*Soviet Canada*) in 1931. This very enthusiastic, readable book, written after the author's journey along the Murmansk Railway, provided a broad view on the development of the region on the basis of technology, scientific exploration, and the heroism of individuals (as opposed to the heroism of the masses, which was becoming a required cliché in propaganda writing). Although published after the end of the decade, it looks now like one of the last examples of 1920s literature of that kind, and likely it was the last published mention of *Soviet Canada*. Already in 1928 the Colonizing Department of the Murmansk Railway experienced severe criticism for its "detachment from the Soviet regime" and lack of a class approach in the selection of colonists, the result of which was the absence of "proletarian settlements" in the area of colonization. Another accusation was that among the new settlers a rather high percentage of people moved within the northern region itself, whereas the main aim was declared as moving people from outside the region.[60]

Figure 5.2 Workers at the fish canning factory belonging to the railway fisheries enterprise Zhelryba, in the Russian town of Kandalaksha, 1926. Courtesy of the Department of Ichthyology and Hydrobiology, St. Petersburg State University (collection of Professor Evgenii Suvorov).

In spite of these criticisms, it is evident that the railway had substantially stimulated in-migration to the region, with a large impact on the growth of Murmansk and several smaller towns. It supported fisheries, especially along the northwest coast of the White Sea, including development of modern fish-processing industry such as canning, which did not exist before in the North (Figure 5.2). However, its own colonization activities remained very modest: only 972 households were resettled, about half of which were the families of railway workers.[61]

Railway economic bodies Zhelles and Zhelryba competed for labour with other large state enterprises in the region, forcing Zhelles to hire even prison labourers for its felling operations. It was, as is argued by Nick Baron, "a policy which significantly contributed ... to the growth of the 'special camps' in the region."[62] This was an answer to the rhetorical question Zorich posed in his book: "What do we need except men's hands now, when all the preliminary work is done and finished, for calling them [the northern regions] to vivid and vigorous life?"[63]

This is an illustration of the utopian vision of many technocratic Russian scientists and officials who thought that scientific exploration and technology would work by themselves to the benefit of the country – that the problem was only to explore natural resources and develop appropriate technology for their exploitation, that people would always be in place, happy to work, and would use all these resources and develop any remote region where these riches were found. And if they didn't, they needed to be taught based on a good working model. If this did not work either, then they could be forced. Following such logic, it is very easy to construct the bridge from the bright utopianism of being a new Canada to the terrible Soviet reality of northern gulags and special settlements. This confirms the existence of tension between utopian vision and dystopian results of colonization policy, which Alfred Rieber attributed to the Russian colonization in general.[64] It is very tragic, however, that such a development of the Soviet economic course of the 1930s fully destroyed many of those who worked for these utopias in the 1920s – Gennadii Chirkin and Vasily Lokot, as well as the head of administration of the Murmansk Railway, Aron Arnol'dov, who also wrote about "railway colonization,"[65] among them. Thousands of other professionals were also accused of being "wreckers," arrested, and killed in prison.

The Murmansk Railway was reorganized and lost its colonizing function in 1930. After that the process of colonization of the North took on well-known features of the Stalinist policy of internal colonialism: forced resettlement of former prosperous peasants from the south (kulaks), collectivization of agricultural and fisheries settlements, rapid industrial development of mining towns, and incredible growth of gulags. The last point in the almost century-long colonization line was the forced removal of all Norwegian and Finnish colonists from the Murman coast in 1940, when under the threat of the coming war it was decided to "clean" the coast of the former "foreign national" residents.[66]

Conclusion

These two cases of unsuccessful attempts to borrow foreign models for colonization of the Russian North have strong similarities, as well as strong differences. The perception of the northern regions as distinct from other parts of the country and the searching for models from countries with similar environments are the most obvious similarities. Understanding of underdevelopment and the backwardness of the Russian North both in terms of institutional, infrastructural developments and the people themselves are also features of both periods under discussion. Yet, although interest to search for new institutional models for colonization was pronounced in pre-revolutionary Russia, especially among the specialists from the Resettlement Administration, under new Soviet power the focus on institutions prevailed. Bolsheviks armed by powerful ideology directed toward creating a New Man did not need any models for individual behaviour. Moreover, local people were no longer considered as the most important users of local resources. The major legacy of the First World War was the strengthening of state power. State became the main resource user, one that could mobilize and move masses on large territories.[67]

The major shift in state policy was made in 1929-30, the period known as "the Great Break," which resulted in rapid industrialization, collectivization of peasants and cultural revolution, and crucial changes in the spatial politics of the country, including the development of the sub-Arctic and Arctic regions.[68] For the first time, the country adopted a clear Arctic policy. The Arctic Ocean was rapidly gaining strategic significance, and the perception of the Arctic and the North in general switched from being "the backs" of the country to becoming an open front: front as a final frontier and front as a place for fighting.[69] Conquering the High Arctic became the goal, and sub-Arctic regions, without losing some of their frontier features, simultaneously became jumping-off places for Arctic battle.[70] However, the northern sub-Arctic regions would lose their specificity when "the economic structure of the country was to become fair and rational, with the ... borderlands no longer colonies of the centre. Dreams of a developed socialist union of republics included ploughing the virgin tundra and breaking the ice of the northern passage."[71]

From 1932, when the ice ship *Sibiriakov* for the first time crossed the Northern Passage in one season, this icy route became the constituent axis for Soviet exploration and mastering the North.[72] Railways remained important, but in popular and propaganda discourses they were overshadowed by icebreakers and aviation. The Main Administration of the Northern Sea Route, organized at the end of the same year, became *imperium in imperio,* absorbing almost everything connected with the North: scientific institutions, trade operations, the mining industry, the "enlightenment" of locals, and, certainly, gulags. In addition, from 1933, when the first ships and submarines of the newly founded Northern Fleet reached the fishing bays on the Murman coast, most of which became closed military cities, the Soviet Arctic policy became firmly connected with militarization. Along with "centralization" and "militarization," the third familiar term from this time was "ideologization," appearing in rhetoric of Soviet heroism in conquering the Arctic.[73] Based on these three tenants, and going its own way in insisting on a sectorial approach to Arctic geopolitics, to scientific exploration, and to building industrial sites, the country found its distinctive northern policy and no longer needed to search for foreign models to master the North.

NOTES

The author would like to thank organizers and participants of the Nordic Environmental History Network workshop "Environmental Histories of the North," 14-15 October 2010, Stockholm, for the most helpful discussion of the first draft of this chapter. I am very grateful to Mary Bailes and Dolly Jørgensen for their kind assistance with language editing.

1 The best discussion of both models of colonization in a broader context of economic transformation of the Kola peninsula is provided by Andy Bruno, "Making Nature Modern: Economic Transformation and Environment in the Soviet North" (PhD diss., University of Illinois, 2011). I am very grateful to Andy Bruno for many discussions we had on the topic since 2003 and for the opportunity to read his dissertation.

On the Norwegian model for colonization, see Alexei Yurchenko and Jens Petter Nielsen, eds., *In the North My Nest Is Made: Studies in the History of the Murman Colonization, 1860-1940* (St. Petersburg: European University at St. Petersburg and University of Tromso, 2005).

On the railway colonization, see the very informative chapter "The Murmansk Railway and Colonization of the North," mostly related to the Karelian part of railway colonization, in Nick Baron, *Soviet Karelia: Politics, Planning and Terror in*

Stalin's Russia, 1920-1939, (London: Routledge, 2007), 75-78, and Olga Kiseleva, "The Activities of the Colonization Industrial-Transport Group of Enterprises of Murmansk Railway in 1923-1929 Years," in *The Industrialisation Process in the Barents Region,* ed. Lars Elenius (Luleå, Sweden: Luleå University of Technology, 2007), 157-66.

On imperial and early Soviet periods of colonization, see Julia Lajus, "Colonization of the Russian North: A Frozen Frontier," in *Cultivating the Colony: Colonial States and Their Environmental Legacies,* ed. Christina Folke Ax et al. (Athens: Ohio University Press, 2011), 164-90; another example of this approach is Ekaterina Orekhova, "Kollonizatsia Murmanskogo poberezh'ia Kol'skogo poluostrova vo vtoroi polovine 19 – pervoi treti 20 vekov" (PhD diss., St. Petersburg State University, 2009).

2 See the published collection by Martin Aust, Ricarda Vulpius, and Alexei Miller, eds., *Imperium Inter pares: Rol' transferov v istorii Rossiiskoi imperii (1700-1917)* (Moscow: Novoe Literaturnoe Obozrenie, 2010).

3 Nicholas B. Breyfogle, Abby Schrader, and William Sunderland, "Russian Colonization: An Introduction," in *Peopling the Russian Periphery: Borderland Colonization in Eurasian History,* ed. Nicholas B. Breyfogle, Abby Schrader, and William Sunderland, 1-18 (London: Routledge, 2007), 7.

4 On the Other, see "Conclusion" in Yuri Slezkine, *Arctic Mirrors: Russia and the Small Peoples of the North* (Ithaca, NY: Cornell University Press, 1994), 387-96.

5 Julia Lajus, Alexei Kraikovski, and Alexei Yurchenko, "The Fisheries of the Russian North, c. 1300-1850," in *A History of the North Atlantic Fisheries,* vol. 1, *From Early Times to the Mid-Nineteenth Century,* ed. David J. Starkey, Jon Th. Thor, and Ingo Heidbrink (Bremen, Germany: Verlag H.M. Hauschild GmbH, 2009), 56-57; Lajus, "Colonization of the Russian North," in Folke Ax, *Cultivating the Colony,* 169-70.

6 Sverker Sörlin, "Rituals and Resources of National History: The North and the Arctic in Swedish Scientific Nationalism," in Bravo and Sörlin, *Narrating the Arctic,* 74.

7 Suzanne Zeller, *Inventing Canada: Early Victorian Science and the Idea of Transcontinental Nation* (Montreal: McGill-Queen's University Press, 2009), 161.

8 Jens Petter Nielsen, "The Murman Coast and Russian Northern Policies ca. 1855-1917," in Yurchenko and Nielsen, *In the North My Nest Is Made,* 24.

9 Julia Lajus, "From Fishing to Mining: The Change of Priorities in the Development of the North and Russian Expedition to Spitsbergen in the Early 20th Century," in *Arktisk Gruvdrift II: Teknik, vetenskap och historia i nor; Foredragpresenterade vid ett seminarium pa Kungl; Vetenskapsakademiem dem 6 maj 2000* (Stockholm: Jernskontorets Berghistoriska Utskott, 2004), 93-106.

10 See, for example, Andy Bruno, "Industrial Life in a Limiting Landscape: An Environmental Interpretation of Stalinist Social Conditions in the Far North," *International Review of Social History* 55 (2010): 153-74.

11 Nikolai Ya. Danilevsky, "Rybnye i zverinye promysly v Belom i Ledovitom moriakh," in *Issledovaniia o sostoianii rybolovstva v Rossii,* vol. 6 (St. Petersburg: Ministerstvo Gosudarstvennykh imuschestv, 1862), 257.

12 Jens Petter Nielsen, "The Murman Coast and Russian Northern Policies ca. 1855-1917" in Yurchenko and Nielsen, *In the North My Nest Is Made,* 13.

13 For a discussion of this issue, see ibid.

14 Ruslan Davydov, "From Correspondence to Settlement: The Colonization of Murman, 1860-1876," in ibid., 29.

15 See Lajus et al., "The Fisheries of the Russian North, c. 1300-1850," 48-50.

16 Ibid., 32.

17 See D.N. Bukharov, *Russkie v Finnmarkene* (St. Petersburg: Tipografiia V. Kirshbauma, 1883); German F. Goebel, "Nasha Severo-Zapadnaia okraina – Laplandia," *Russkoe sudokhodstvo* no. 10-12 (1904); no. 1-4, 6-8, 10-11 (1905); Leonid L. Breitfuss, *Rybnyi promysel russkikh pomorov v Severnom Ledovitom okeane; ego proshloe i nastoiaschee* (St. Petersburg, 1913).

18 Nikolai Ya. Danilevsky, *Rossia i Evropa: vzgliad na kul'turnye i politicheskie otnosheniia Slavianskogo mira k Germano-Romanskomu* (St. Petersburg: Izdanie Tovarishchestva tovarishchestva obshchestvennaia pol'za, 1871).

19 Danilevsky, "Rybnye i zverinye promysly," 243-44.

20 Walter Sperling, "Building a Railway, Creating Imperial Space: 'Locality,' 'Region,' 'Russia,' 'Empire' as Political Arguments in Port-Reform Russia," *Ab Imperio* 2 (2006): 101-34.

21 Ibid., 121.

22 Alexander Engelhardt, *Russkii Sever* (Arkhangelsk, 1897); Henry Cooke, trans., A *Russian Province of the North* (Westminster [London]: Archibald Constable, 1899).

23 Helen Rozwadowsky, *The Sea Knows No Boundaries: A Century of Marine Science under ICES* (Copenhagen: ICES with University of Washington Press, 2002).

24 For the Swedish tradition of using natural science and statecraft to colonize the North, see Sörlin, "Rituals and Resources of Natural History."

25 See, for example, Nikolai A. Varpakhovsky, ed., *S'ezd russkikh rybopromyshlennikov* (St. Petersburg, 1889); Mikhail F. Mets, "Neobkhodimost' nauchnopromyslovykh morskikh issledovanii u beregov Murmana," in *Trudy Severnoi komissii 1897-1898 gg.* (St. Petersburg, 1898), Suppl. 1, 1- 4; German F. Goebel, *O nuzhdakh Murmana* (St. Petersburg, 1897).

26 See Sörlin, "Rituals and Resources of Natural History"; and Zeller, *Inventing Canada.*

27 See *Trudy Severnoi komissii 1897–1898 gg.,* 1-4.

28 For more details on the Murman expedition, see Julia A. Lajus, "'Foreign Science' in a Russian Context: Murman Scientific-Fishery Expedition and Russian Participation in Early ICES Activity," *ICES Marine Science Symposia* 215 (2002): 64-72.

29 See Nikolai A. Varpakhovsky, "O Nauchno-promyslovykh morskikh issledovaniiakn zagranitsei," in *Trudy Severnoi komissii 1897-1898 gg.,* 31-36.

30 See Alexei E. Karimov, *Doku da topor i sokha khodili: Ocherki istorii zemelnogo i lesnogo kadastra v Rossii XVI – nachala XX veka* (Moscow: Nauka, 2007), 182-84.

31 See Konstantin Yu. Spade, "Severnye rybnye promysly i neotlozhnye mery k ikh razvitiiu," *Izvestiia Arkhangelskogo Obschestva Izucheniia Russkogo Severa* 21 (1911): 716-24.

32 These discussions were reflected on the pages of the journal *Izvestiia Arkhangelskogo Obschestva Izucheniia Russkogo Severa* from 1908 to 1918.

33 For the history of Soviet railways, see E.A. Rees, *Stalinism and Soviet Rail Transport, 1928-41* (New York: Basingstoke, 1995); on the important role of railways in the early years of the Soviet economy, see Anthony Heywood, *Modernising Lenin's Russia: Economic Reconstruction, Foreign Trade and the Railways* (Cambridge: Cambridge University Press, 1999).

34 The word "combinat" is used in both German and Russian; for the full definition of Soviet combinates, see Bolshaia Sovetskaia Entsiklopedia 1969-78, http://slovari.yandex.ru/.

35 Gennadii F. Chirkin, *Sovetskaia Kanada* (Karelo-Murmanskii krai) (Leningrad: P.P. Soikin, 1929), 3, 8; on the details of this decree, see Baron, *Soviet Karelia,* 77.

36 Ibid., 9.

37 Chirkin, *Sovetskaia Kanada,* 1. For a recent discussion on the role of infrastructure in shaping the Canadian nation, see Jonathan F. Vance, *Building Canada: People and Projects That Shaped the Nation* (Toronto: Penguin, 2006).

38 See Wynn, *Canada and Arctic North America,* 397.

39 For the history of Canada Pacific Railways, see Pierre Berton, *The Impossible Railway: The Building of the Canadian Pacific* (New York: Alfred A. Knopf, 1972) and the classic work by Harold A. Innis, *A History of the Pacific Railway* (London: P.S. King & Son, 1923). For land and colonization policies of this railway, see James B. Hedges, *Building the Canadian West – The Land and Colonization Policies of the Canadian Pacific Railway* (New York: Macmillan, 1939); chap. 14, "The Lands," in J. Lorne McDougall, *Canadian Pacific: A Brief History* (Montreal: McGill-Queen's University Press, 1968), 125-42.

40 See T.D. Regehr, *The Canadian Northern Railway: Pioneer Road of the Northern Prairies, 1895-1918* (Toronto: Macmillan, 1976).

41 Chirkin, *Sovetskaia Kanada,* 4.

42 Gennadii F. Chirkin, *Transportno-promyshlennyi kolonizatsionnyi kombinat Murmanskoi zheleznoi dorogi: ego vozniknovenie, razvitie i metod rabot* (Moscow and Leningrad: Glavnauka, 1928)/Trudy GosNII zemleustroistva i pereselenia, vol. 9, 23.

43 Wynn, *Canada and Arctic North America,* 227.

44 Willard Sunderland fairly referred to this administration as one of "the primarily colonial institution" for Russia; see Willard Sunderland, "Ministerstvo Aziatskoi Rossii: Nikogda ne suschestvovavshee, no imevshee dlia etogo vse shansy kolonialnoe vedomstvo," in Aust, Vulpius, and Miller, *Imperium Inter pares,* 105-49.

45 Peter Holquist, "'In Accord with State's Interests and the People's Wishes': The Technocratic Ideology of Imperial Russia's Resettlement Administration," *Slavic Review* 69, 1 (2010): 152.

46 See Benjamin Fritjof Schenk, "Imperial Inter-Rail: Vlianie mezhnatsional'nogo i mezhimperskogo vospriiatiia i sopernichestva na politiku zhaleznodorozhnogo stroitel'stva v tsarskoi Rossii," in Aust, Vulpius, and Miller, *Imperium Inter pares,* 354-80.

47 Ibid., 369-70.

48 On Chirkin's earlier writing, see, for example, Gennadii F. Chirkin, "Kolonizatsionnoe i narodno-khoziaistvennoe znachenie proektiruemoi Yuzhno-Sibirskoi magistrali"

in *Voprosy kolonizatsii: Periodicheskii sbornik,* ed. G.F. Chirkin and N.A. Gavrilov (Petrograd, 1913): 100-28; and "O kolonizatsionno-kulturnykh zadachakh posle voiny," in *Voprosy kolonizatisii: Periodicheskii sbornik,* vol. 18, ed. G.F. Chirkin and N.A. Gavriov (Petrograd, 1915), 25-49.
49 Gennadii F. Chirkin, *Kolonizatsiia Severa i puti soobscheniia* (Petrograd, 1920), 19.
50 Holquist, "In Accord with State's Interests," 155.
51 See Schenk, "Imperial Inter-Rail," 369.
52 The term "settlement" instead of "colonization" was in use during the Soviet era. See, for example, the preface to a book by British geographer Terence Armstrong, *Russian Settlement in the North* (Cambridge: Cambridge University Press, 1965), where he wrote about colonization that "it is as part of the history of man's relation to his environment that this story is mainly interesting" (10).
53 A. Zorich, *Sovetskaia Kanada: Ocherki* (Moscow: Izd. "Federatsia," 1931), 35-36.
54 Baron, *Soviet Karelia,* 77.
55 Ibid., 34.
56 Chirkin, *Sovetskaia Kanada,* 5.
57 Chirkin, *Transportno-promyshlennyi kolonizatsionnyi kombinat,* 28-29.
58 See John McCannon, *Red Arctic: Polar Exploration and the Myth of the North in the Soviet Union* (New York and Oxford: Oxford University Press, 1998); Alla Bolotova, "Colonization of Nature in the Soviet Union: State Ideology, Public Discourse, and the Experience of Geologists," *Historical Social Research* 29, 3 (2004): 104-23.
59 Zorich, *Sovetskaia Kanada,* 132.
60 Baron, *Soviet Karelia,* 78.
61 Ibid.
62 Ibid.
63 Zorich, *Sovetskaia Kanada,* 75, emphasis in original.
64 Alfred J. Rieber, "Colonizing Eurasia," in Breyfogle, Schrader, and Sunderland, *Peopling the Russian Periphery,* 267.
65 Aron Arnol'dov, *Zheleznodorozhnaia kolonizatsiia v Karelsko-Murmanskom krae: Po materialam razrabotannym Kolonizatsionnym otdelom pravlenia dorogi* (Leningrad: Pravlenie Murmanskoi zheleznoi dorogi, 1925).
66 See details in Alexander Portsel, "The Norwegian Colonists in Murman during the Soviet Period (1920-1940)," in Yurchenko and Nielsen, *In the North My Nest Is Made,* 158.
67 I develop this argument elsewhere; see Julia Lajus, "Expertise, Governance and the Marginalization of Local Users in Fisheries," under submission to *Environment and History.* In general, very important analyses of changes in "population" politics is provided in works by Peter Holquist, especially "To Count, to Extract, and to Exterminate: Population Statistics and Population Politics in Late Imperial and Soviet Russia," in *A State of Nations: Empire and Nation-Making in the Age of Lenin and Stalin,* ed. Ronald Suny and Terry Matin, 111-44 (Oxford: Oxford University Press, 2001).
68 On these changes in spatial politics, see "Totalitarianism and Space," in Baron, *Soviet Karelia,* 240-42.

69 Both fighting "with the Arctic" and fighting "for the Arctic" are well reflected in titles of many popular books and media publications; in English see, for example, Harry P. Smolka, *40,000 against the Arctic: Russia's Polar Empire* (New York: William Morrow, 1937).

70 As I argue elsewhere, "strong economic, environmental, and security constraints impede the development of the Russian northern frontier, which is likely to remain 'frozen'"; see Lajus, "Colonization of the Russian North," in Folke Ax et al., *Cultivating the Colony,* 184. The same metaphor of "frozen frontier" is independently used in the title of the most recent book by Alan Wood, *Russian's Frozen Frontier: A History of Siberia and the Russian Far East, 1581-1991* (New York: Bloomsbury Academic, 2011).

71 Slezkine, *Arctic Mirrors,* 265.

72 See Linda Trautman, "Modernization of Russia's Last Frontier: The Arctic and the Northern Sea Route from the 1930s to the 1990s," in *Modernization in Russia since 1900,* ed. Markku Kangaspuro and Jeremy Smith (Helsinki: Finnish Literature Society, 2006), 252-66.

73 See McCannon, *Red Arctic;* and Pier Horensma, *The Soviet Arctic* (London: Routledge, 1991), esp. 57-60.

PART 3

WORKING THE NORTH

Traversal Technology Transfer
The Transfer of Agricultural Knowledge between Peripheries in the North

JAN KUNNAS

During the nineteenth century, agricultural knowledge was widely shared and distributed in Europe. This transfer often took place from scientifically developed countries like Germany or Great Britain to technologically less-developed countries like Sweden. For example, many German and English agricultural textbooks were translated to Swedish, such as the works by Englishmen James Caird, Henry Stephens, and George Stephens, and by Germans Justus von Liebig and Albrecht Thaer.[1] Furthermore, works on the agriculture of Denmark, the Netherlands, Belgium, France, and Ireland were translated to Swedish.[2]

In this chapter, I argue, however, that technology transfer was not a one-way process going solely from the interior to the periphery. I show that there was also ongoing traversal technology transfer, from the periphery to the periphery. A traversal approach like this not only brings a more diversified view on historical processes shaping the North but also gives insights to the current discussion on technology transfer as a means of development. My argument is that the suitability of a technology to the particular nexus of social and natural conditions it is deployed in is more important than how advanced it is.

This chapter uses two examples centred on Sweden to demonstrate traversal technology transfer and its environmental effects. My first example is the transfer of Finnish slash-and-burn cultivation know-how in the sixteenth and seventeenth centuries to mid-Sweden (1 on Map 6.1), thence further up

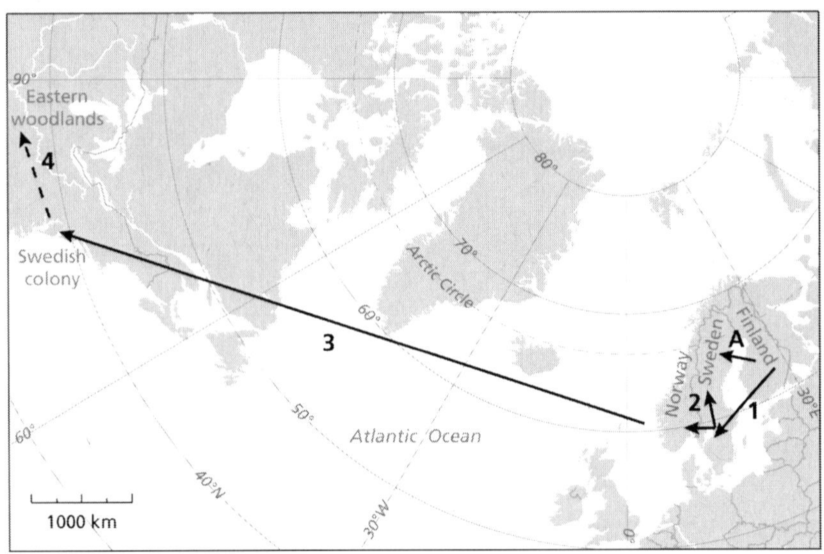

Map 6.1 The transfer of slash-and-burn and peatland cultivation know-how

north and to the forest region between Sweden and Norway (2 on the map), and finally through the Swedish colony in North America (3 on the map), and to the eastern woodlands (4 on the map). My second example is the transfer of peatland cultivation practices across the Gulf of Bothnia from Ostrobothnia to Västerbotten in the eighteenth and nineteenth centuries (A on Map 6.1). The later transfer operated on a much smaller geographical scale, which does not, however, diminish its importance. Indeed, the importance of a transfer should not be judged by the distance of the transfer but by the impact in the recipient area.

These two cases can be considered as technology transfer from the periphery to the periphery, as all the regions included can be considered as belonging to the periphery on a European or global scale. Both Sweden and Finland are small countries at the northern fringe of Europe. The Swedish population in 1700 was 1.3 million (1.5 percent of the total population of Europe) and 2.6 million (1.9 percent) in 1820. The population of Finland was in 1700 about 400,000 or 0.5 percent of the total population of Europe, and despite a tripling of the population to 1.2 million, the share in 1820 was still only 0.9 percent.[3] Finland is for our purposes here treated as a separate entity – even though it was a part of the Swedish Empire until 1809, and then a part of the Russian Empire until 1917 – since Finland was to some degree treated as a separate entity within both empires and had some degree of autonomy.

The peripheral location in the North also meant vulnerability to climatic conditions. In Finland, harvest failure attributed to summer frosts, damp weather, and short growing seasons resulted in the death of 23 to 33 percent of the population in 1695-97 and is blamed for the premature death of 7 percent of the population in 1866-68. Sweden also faced similar weather disturbances, but in the 1860s only one Swedish province, Norrbotten, faced severe famine.[4]

Technology and Its Transfer

Both "technology" and its "transfer" have been given such a wide variety of meanings that it is useful to specify how I use these terms in the case studies. "Technology transfer" can be defined as a broad set of processes covering the flows of know-how, experience, and equipment from one place, group, or person to another. It is usually described as something going from developed countries to developing countries, or from the interior to the periphery, or from "advanced" to "backward" areas.[5] This chapter takes a critical perspective on this conventional view with its two pre-industrial examples of traverse technology transfer – from the periphery to the periphery.

The narrow definition of technology as the application of scientific theory to the solution of practical problems is extensively criticized.[6] Historian of technology Thomas P. Hughes sees technology as "craftsmen, mechanics, inventors, engineers, designers, and scientists using tools, machines, and knowledge to create and control a human-built world consisting of artifacts and systems associated with ... engineering [and] ... used as a tool and as a source of symbols by many architects and artists."[7] Another historian of technology, Thomas J. Misa, abandons the quest for a definition, arguing that the insight that technological changes generate a complex mix of economic, social, and cultural changes is more valuable than having a simple definition of a complex term.[8]

In this chapter, "technology" refers to new and better ways of achieving economic ends that contribute to economic growth.[9] Whether or not the technology used originates from scientific theory does not matter. My focus is on agriculture, thus technology means here the ability to feed more people (extensive growth) or provide a better level of nutrition (intensive growth). By "technology transfer" I mean the "utilization of an existing technique in an instance where it has not previously been used."[10]

Slash-and-Burn Cultivation

Slash-and-burn cultivation can be defined as a cultivation method of cutting living trees to clear land, burning the biomass after letting it dry, and planting a crop in the ashes in an appropriate season. In forested areas, slash-and-burn agriculture was generally the first method used to cultivate grains and is thus usually described as a primitive cultivation method. Compared with the more "advanced" field cultivation based on external fertilizers, it has, however, several major advantages. First, there is no need for external capital or any large investments. This is partly because neither fertilizers (besides ash) nor a great number of draft animals are necessarily needed. One horse per farm was generally enough. Second, it required less work than clearing land for permanent cultivation. This difference between the amount of work needed to clear a field and the amount needed in slash-and-burn cultivation was emphasized in eastern Finland, where the soil is mainly glacial till and therefore stony.[11]

Over the centuries, slash-and-burn cultivation technology developed, and different variants emerged according to the type of forest it was used in. For this story of technology transfer, we are interested in a particular method called *huuhta,* which was likely developed in the Middle Ages by the Vepsians (a Finno-Ugric people) or the Savonians (a tribe of eastern Finland). Together with a new grain variety *korpiruis* (clearing rye), this particular method made it possible to practise slash-and-burn-cultivation in forests dominated by spruce (*Picea abies*). In spruce-dominated forests, it was superior to the less advanced *pykälikkömaa* (notched land) method, which was aimed first at converting coniferous forests into deciduous ones through ring-barking the spruces or pines. The ring-barked trees were left to dry for twenty years on average, during which time broad-leaved trees grew between the dying trunks. This was obviously a time-consuming procedure and resorting to that would have slowed down the colonization processes of the forested areas of Finland and Sweden.[12]

The new rye variety could effectively utilize the nitrogen in the ash, and a typical harvest was about twenty grains to every grain sown. In favourable conditions, thirty- to forty-fold yields were not uncommon, whereas for ordinary field rye, the corresponding ratio between harvest and seed was about ten to one. Powered with this cultivation innovation and new grain variety, the Savonian tribe settled most of the north-central provinces of Finland in a relatively short time. In the eighteenth century, swiddened areas in eastern Finland even acted as a grain source for Ostrobothnia, in western

Finland, where the cultivation of permanent fields and tar production were practised instead of swiddening. Still, as late as the 1830s, in the eastern and central parts of the country, every third loaf of rye bread was baked of grain from swidden harvests (Figure 6.1).[13]

Figure 6.1 The high yield of clearing rye is illustrated by this exhibit in the Skräddrabo Finnskogsmuseet. This is the yield from thirty-five grains. Photo by the author.

Around 1580, Savo-Karelian settlers started to arrive in the forest areas of middle Sweden, primarily the dukedom of Duke Charles, west of Stockholm. The role of the duke in this movement of people has been discussed widely. I agree with the observation by the Finnish historian Kari Tarkiainen that it is difficult to accept a totally spontaneous movement over a distance of a thousand kilometres to the dukedom in a time where official permission was needed to move. When Duke Charles became regent in 1592, and finally King Charles IX of Sweden in 1604, the forests in northern Sweden were also opened for Savo-Karelian settlers, who eventually settled all the way up to southern Lapland.[14]

These settlers brought with them a whole technological package built around slash-and-burn cultivation. Apart from the swidden rye, the settlers brought also a special variety of turnip (*Brassica rapa* var. *rapa*), the so-called swidden turnip (*kaskinauris*), which was cultivated instead of or along with swidden rye. They brought a plow (*finnplog*) that was especially suitable for cultivation on the stony soil of the Finnbygden, and the East Finnish cow, considered particularly appropriate for woodland areas. Furthermore, they brought along their traditional set of buildings, including the smoke house (*savupirtti*) and drying barn (*riihi*), as well as tools, furniture, hunting weapons, and fishing implements.[15]

The effectiveness of the Finnish cultivation method, primitive or not, is shown clearly by the fact that the households in areas inhabited by people of Finnish origin (*finnbygderna*) in Hälsingland paid in general around double the amount of taxes as the households in central districts. Thus, it was hardly surprising that the Finnish slash-and-burn cultivation methods were over time disseminated to farmers of Swedish origin, who gladly swiddened on their outer fields (*utmark*). According to museum director Kjell Lööw, the Finns and the Swedes cooperated: the Swedes contributed with land and the Finns brought their know-how. Such cooperation was so common that in the 1664 forest ordinance there was a special paragraph against this procedure.[16]

The very source of this success, the high-yielding but extremely wood-consuming cultivation method, caused a sudden end to the swidden honeymoon. A large wood consumption was no problem as long as there were plentiful new forest areas to use, but the rise of copper and iron industries in the Bergslagen district in middle Sweden, which turned to the same forests for their ever increasing need of wood, caused conflicts. Welcomed settlers became forest-destroying criminals during the seventeenth century. Several laws were passed in an attempt to curtail the Forest Finns' forest use, and their expulsion from the Crown forests was demanded by the governor of the Bergslagen district, Gustaf Leijonhufvud, and others representing the interests of the industries.[17] As a result, the Finnish settlers needed to find new areas to settle. First, they moved further west, then eventually to eastern Norway, where the first Forest Finns settled permanently during the 1620s.[18] Second, the Swedish government had in 1638 founded a colony on the Delaware River, in North America, and the Crown had difficulty finding eager emigrants. It thus began sending "forest-destroying Finns" there to solve two problems at once. Finns sentenced for illegal slash-and-burn cultivation were pardoned to go to New Sweden with their families, while others went voluntarily.[19] This "voluntarism" was strengthened by Queen Christina's order to burn down the huts and houses of the swiddening Finns causing great damage to the mining industry, and to destroy their crops, so that they would be forced to leave the forests.[20]

The immigration of Finns to North America from Sweden, and increasingly straight from Finland, did not stop when the settlement was seized by the Dutch and incorporated into New Netherland in 1655, nor with the English conquest of New Netherland in 1664. Compared with the total number of immigrants arriving to North America, the Finns were a minuscule part, but they brought with them a whole technological package, which was to play a big role in the settlement process of the continent. Some parts of the package did not fit into the cultural and environmental circumstances of the new surroundings and were replaced by indigenous knowledge learned from the Aboriginals. For example, rye was replaced by Indian corn. This learning process was facilitated by a similarity of lifestyles and the Finns' previous cultural interaction with the indigenous Sami population, which made the Finns culturally adaptable to life on the Indian frontier. The surviving Finnish technological imports included a traditional farmhouse found particularly in Finland, which became the prototype for the Midland American dogtrot building, and the worm fence, which the settlers might have learned from the Sami in the first place.[21]

Most important was, though, their slash-and-burn cultivation know-how, which eventually diffused to settlers of different origin. US geographers Terry G. Jordan and Matti Kaups propose that seventeenth-century Finnish settlers of the New Sweden colony on the Delaware River made fundamental contributions to the successful and rapid occupancy of the eastern woodlands of North America through their initiation of large-scale forest removal, which opened the wedge for permanent settlement of the land. They argue that this successful forest pioneering culture permitted the United States to become a transcontinental nation rather than remaining a littoral state clinging to the Atlantic. If their suggestion is right, then indeed the second half of US historian Frederick Jackson Turner's claim regarding early American history that "too exclusive attention has been to the German origins, too little to the American factors" should be rewritten as "too little to the American factors and Finnish origins."[22]

The speeding up of the occupancy process, first in the forest between Sweden and Norway, and later – and most importantly – in the eastern woodlands of North America, makes slash-and-burn cultivation technology the single most important technology coming from Finland so far when measured by global impacts.

Burning Cultivation of Peatlands

One advantage of slash-and-burn cultivation is that no fertilizer is needed besides the ash from the trees burned. In field cultivation, on the contrary, the availability of manure for fertilizing was a constant problem. The rising ambition to break up new land for cultivation made the constant manure shortage extremely acute in Sweden at the end of the eighteenth and beginning of the nineteenth centuries.[23] One solution to this was the use of fire, but this time on grass- and peatlands instead of forests. According to historian of fire Stephen J. Pyne, two strategies emerged: "One imitated slash-and-burn forestry in which paring, drying, and burning occurred more or less on the site. The other mimicked charcoal, tar or potash production, in which the cut turf was gathered into piles, often covered, sometimes placed into ovens, and slowly burned; peasants then carried the ashes back to the fields."[24] According to Pyne, these practices achieved greatest renown in Britain, which had upland bogs in abundance.

Etnologist Nils-Arvid Bringéus gives Jacob Serenius and his book *Engelska åkermannen och fåraherden* (The English field man and shepherd) published in 1727 the honour of being the first presentation to Swedish readers of British burning cultivation methods and the breast-plow used to pare the turf. The first picture of the breast-plow was published in the book *Sten Biörnsons jordmärg, eller trogna bondaläro,* in 1745.[25] Although the writer used the pseudonym Sten Biörnsons, he was, in fact, Stephen Bennet, who was born in Reasby, Leicester. Inspiration for burning cultivation was also gained through translations of German textbooks, such as Albrecht Thaer's *Grundsätze der rationellen Landwirthschaft,* which was published in Swedish in 1817 and reprinted in 1846 and 1861.[26]

Although these texts may have encouraged burning cultivation among a certain class of readers, a more likely source for inspiration for the northern parts of Sweden was the agricultural practice in Finnish Ostrobothnia, on the other side of the Gulf of Bothnia. Finland was part of Sweden until 1809, and climatic conditions were similar on both sides of the gulf. The English example was important, however, as its agriculture was considered the best imaginable.[27] Therefore, the practice of burning cultivation in England also worked as an endorsement for it in Sweden, even though it was practised with Finnish methods.

Native Finns described the Finnish methods in Swedish by the late eighteenth century: Professor Pehr Adrian Gadd's *Inledning till Svenska Landt-Skötseln* (1773) and Bishop Jacob Tengström's *Försök till lärobok i landthushållningen för finska bonden* (1803) both had detailed descriptions about Finnish burning cultivation of peatlands.[28] The text box presents an outline of Tengström's description of burning cultivation. This description clearly shows that it was by then a well-developed cultivation method, with several sequential measures.

That burning cultivation of peatlands had developed into such an advanced multi-phase cultivation method is no coincidence considering that it probably has a very long history in Finland. Finnish historian K.R. Melander argues that it is possible that burning cultivation of peatlands was already practised in Finland at the beginning of the fourteenth century. Then, however, fire was probably used to create pasture, as it did not require as thorough drainage as did grain cultivation.[29] Urban Hiärne mentions the priest in Storkyro in Ostrobothnia, Isak Brenner, as the inventor of burning cultivation of peatlands in Finland.[30] Yet, Brenner got his first yields from burned peatlands in the 1660s, whereas court protocols show that peatlands were already being burned in 1640.[31] This reminds us about a general problem in historical research: that written sources usually highlight powerful men, while the common people usually remain in the shadows. Nevertheless, Brenner's good harvests contributed to the dissemination of burning cultivation of peatlands in southern Ostrobothnia. Similarly, in north Karelia burning cultivation was spread by the encouraging examples of the vicars Jacob Stenius the Older (1704-66) and Gabriel Wallenius (1725-1808).[32] Finnish geographer Helmer Smeds argues that burning cultivation of peatlands had arisen endemically by the mid-seventeenth century from the use of fire in order to create pastures, or that it was by then already an old and well-known practice.[33] According to a description by Mikael Wexionius

Tengström's description of burning cultivation of peatlands at the beginning of the nineteenth century

First year

1. The drain ditch is dug 1.8 metres wide and at least 1.2 metres deep.
2. The surface drain ditches (*laggdiken*) are dug 1.2 metres wide and 0.9 metres deep.
3. The feeder drain ditches (*skärdiken,* or *tegdiken*) are dug at a minimum distance of 36 metres from each other.

Second year

4. The swamp will rest a year or more, though usually only one.

Third year

5. Thirty-six-metre-wide strips divided by the feeder drains are divided by ditches into three parts of equal size.
6. The forest is cleared, and stumps and roots dug up.
7. The stumps and wood are dragged to the farm during winter.

Fourth year

8. The swamp is hoed with large hoes (22 cm high and 15 cm wide).

Fifth year

9. The swamp is plowed with an earthing-up plow (*årder*) and harrowed (*harvning*).
10. The plowing and harrowing is repeated.
11. The swamp and twigs are burned.
12. The swamp is harrowed.
13. The swamp is burned.
14. The swamp is harrowed.
15. The swamp is burned once more, if possible.
16. Rye, ideally root rye, is sowed before mid-August.
17. The grain is dug in with a twig harrow (*kvistharv*), a harrow made of split tree trunks with the branches intact.

Sixth year

18. The rye is harvested.

Seventh year

19. Plowing begins in the spring, once the swamp starts to melt (*källossningstiden*).
20. The swamp is harrowed.
21. The swamp is burned several times and always after harrowing.
22. Rye is sowed.

Eighth year

23. The rye is harvested.

Source: Jacob Tengström, *Försök till lärobok i landthushållningen, för finska bonden* (Åbo, 1803).

from 1650, the landscape Savolax in eastern Finland got its name from the smoke (*savu*), which filled every corner and bay, from all the forests and swamps the inhabitants had ignited.[34]

According to Finnish agronomist Gösta Grotenfelt, peatland cultivation was relatively widespread in several parts of the country during the last decades of the eighteenth century. In the 1780s and 1790s, it was practised as far north as the polar circle in Rovaniemi by Vicar E. Fellman and Lieutenant M.F. Clementoff.[35] It was, however, most common in the peat-rich region of Ostrobothnia, on the western coast of Finland, where it compensated for the diminution of forest resources caused by slash-and-burn cultivation and tar burning.[36] In some districts of Ostrobothnia during the 1820s and 1830s, over half the yield of some grains came from peatland cultivation.[37]

During the latter half of the 1810s, the Finnish cultivation method of chopping and burning was introduced in northern Uppland, where it replaced the more laborious and thus also more expensive methods using plows (*skumplöjning*) or shovels.[38] Different cultivation methods could also complement each other; in 1819, both chopping and plowing followed by burning were used on the properties of the Dannemora mine for the clearance of new fields from bogs and moss, and at the properties of the Överby factory for the cultivation of meadows growing mosses (*mossvall*).[39] The direction for the sources of influence can as well be seen in the regional variation of the tools used: the paring spade or breast-plow in southwest Sweden, and the hoe elsewhere.[40]

As proof of the continuous transfer of peatland cultivation practices across the Gulf of Bothnia, we have accounts of local descriptions of attempts to cultivate peatlands using Finnish cultivation methods. As an example, there is a description from 1827 of cultivation trials in Lövånger parish in Västerbotten:

> Several experiments have been done with the cultivation of mosses and tufty soils that had been chopped up and burned according to the Finnish method. Such cultivation is considered less permanent; but it is possible, that the choice of soil and their treatment requested special knowledge, and would thus be desirable that a skilled and active farmer could in Finland acquire the knowledge, after scientific bases, about this cultivation method and spread it to locations that have so far lacked the knowledge, as it is not likely to be disproved, that Finland's inhabitants, through wise treated soil burning (kyttning), have got most of their abundant grain yield, which has

> there created general well-being, and which should not with similar soil use be for Västerbotten less strange.[41]

The good reputation of Finnish seed in Sweden encouraged the Finnish cultivation method as well, to maintain the seed quality.[42] Swede Magnus Stridzberg, for example, praised the Finnish rye seed in 1727, claiming that the coarse rye brought from Finland and coming from the burned bogs and swamps gave the first year a double yield compared with the yield from common rye, but the yield diminished the following year and so forth, until it eventually did not differ from common rye.[43]

Fireless Cultivation

Along with Finnish agricultural technology based on slash-and-burn cultivation and burning cultivation of peatlands, there was also Finnish know-how of fireless methods. For example, a Finnish method for the transformation of bogs into meadows through freezing gained Swedish interest as well. This interest is exemplified by a translation of a Swedish text from 1865:

> Bogs, that are unsuitable for field cultivation, can be changed to meadows, if one can on occasion set them under water, the water should remain over the winter and freeze to the bottom, thereafter the water should remain at least at a half aln [0.3 metres] above the moss, until it starts to rot because of the lack of air and the heat of the water and starts to transform into a watery mud, then the water is slowly released in its entirety, so that the soil in the water can settle down on the dead and rotting moss, which instead of growing changes into mould, making growth of grasses possible. This cultivation method has since old times been practiced in Finland ... a similar method with submersion of the mosses has lately been used in Västerbotten and several other northern provinces, and this most peculiar cultivation method has made it possible to transform otherwise useless peat-mosses [*hvitmossar*] into meadows.[44]

Burning cultivation was not favoured out of some kind of fascination with fire but because it was the most labour-effective method to clear and prepare forests and peatlands for cultivation. Slash-and-burn cultivation was eventually abandoned in Finland as more valuable uses of the forests appeared with Finland's wood-based industrialization in the late nineteenth century. It was, though, replaced by burning cultivation of peatlands, which peaked in the 1890s. As cultivation methods replacing fire with the use of sand, clay, and fertilizers were developed, burning cultivation of peatlands faded out by 1940.[45]

Conclusion

Technology transfer is normally described as a one-sided flow from the centres to the periphery. In this chapter, I have shown that it is a much more multi-faceted phenomenon with examples on a technology transfer between the peripheries in the North. In all these cases, the usefulness and suitability of the technology was more important than how sophisticated it is considered.[46]

The North is often defined through extremes, such as the cruelty of the climate or untamed wilderness, extremes for which technology can be used to protect humans from it or elevate them above it. Technology originating in the North had to deal with these extremes. The northern agricultural technologies of slash-and-burn and peatland burning were well adapted to the surrounding environment – soil types, climate, and ecosystems – which enhanced its traversal transfer from one northern peripheral region to another.

The northern populations were not just the product of their environment; on the contrary, they actively made their northern environments. Whether this adaptation of their environment to their own needs was for good or bad is another question. The burning cultivation of peatlands was by far the greatest source of carbon dioxide in Finland throughout the nineteenth century and at the beginning of the twentieth century; this might also apply for Sweden and other countries where burning cultivation of peatland was practised. Old swidden areas have been reforested, but the marks of burning cultivation of peatlands are still clearly visible in the Finnish landscape as large field plains, especially in southern Ostrobothnia, which are nowadays considered valuable landscapes. Several species dependent on burned forests or broad-leaved forests following slash-and-burn cultivation are now classified as threatened.[47]

The regions under scrutiny might have been peripheral in a European perspective, but that did not mean they were isolated and passive recipients of ideas and technology coming from the south. Rather, they were active developers of technology on their own, as well as open to new ideas, whatever direction they came from. The inspiration and endorsement for burning cultivation of peatlands might have come from English and German textbooks and example, but the technology used travelled in a traversal direction from the Finnish periphery to the Swedish periphery. The northern origin was even clearer regarding slash-and-burn cultivation technology.

The technology transfer processes for slash-and-burn cultivation and burning cultivation of peatlands had several similarities but also one major difference. The slash-and-burn cultivation technology was incorporated into a comprehensive technological and cultural package. This package included tools, plants, building techniques, and collateral livelihoods, such as animal husbandry on the pastures created through the burning. Of paramount importance was the culture formed around these livelihoods. In the case of burning cultivation of peatlands, only a simple technological package consisting of the know-how of cultivation and the tools needed was transferred. This difference has large cultural implications: whereas the legacy of the cultural system built around slash-and-burn cultivation is still alive, burning cultivation of peatlands is completely forgotten. Interestingly, the memory of the former is more visible in the receiving area – that is, the Finnskog area – in present-day Sweden, with special museums like Torsby Finnkulturcentrum and the Finnskogsmuseet in Skräddrabo built around the legacy of the Forest Finns, than in their place of origin in eastern Finland.[48] The Forest Finns' former smoke huts are among the most visible remains of their culture; it is also one of the least advanced parts of their technological package, thus downplaying the more advanced elements of this culture.

Similar traversal technology transfer along the latitudes likely happened at all latitudes, not only in the North, although researchers have yet to investigate these transfers. They were most likely done through the movement of people and their ideas, and through a transfer of tools and plants – usually a combination of these different carriers – with written documents playing only a secondary role. The transfer process could include either comprehensive technological and cultural packages, or simple technological packages. These multi-faceted processes of technology transfer pose great challenges to historical research that is mostly based on written documents and is likely to cause an overemphasis on the technology transfer from the centres to the periphery and an underappreciation of the transfer between the peripheries. The effort needed to track the technology transfer between the peripheries is, however, rewarding: not only does it bring a more diversified view of historical processes but it can also provide insights to the current discussion on technology transfer as a mean of development.

NOTES

1 James Caird, *English Agriculture in 1850-51* (London: Longman, Brown, Green, and Longmans, 1852). Swedish translation: *Englands åkerbruk åren 1850-51* (Örebro: N.M. Lindh, 1855); Henry Stephens, *The Book of the Farm*, vols. 1 to 3 (Edinburgh and London: William Blackwood and Sons, 1844). Swedish translation: *Landtbrukets bok* (Stockholm, 1856-61); George Stephens, *Essay on the Utility, Formation and Management of Irrigated Meadows* (Edinburgh: 1826). Swedish translation: *Afhandling om ängsvattning, dikning och vallars anläggning*, 1841. Justus von Liebig, *Naturvetenskapliga bref öfver vår tids landtbruk* (Örebro, 1861); *Chemien tillämpad på jordbruk och fysiologi* (Stockholm, 1846); *Åkerbruks-kemiens grundsatser med hänseende till de i England utförda undersökningar* (Stockholm, 1856); *Naturlagarna för växtnäringen eller kemien tillämpad på landthushållning och fysiologi* (Upsala, 1867); *Naturlagarna för åkerbruket* (Örebro, 1864); Albrecht Thaer, *Den rationela landthushållningens grundsatser*, n. 1-4 (Stockholm, 1816-17); *Inledning till kunskapen om engelska landthushållningen ...* (Götheborg, 1801); *Landtbruksbokhålleri* (Stockholm, 1816).

2 H.J.G.A. Lacoppidan, *Landthushållningslära: Företrädesvis för mindre landtbruk* (Stockholm: Sigfrid Flodins Förlag, 1869); Émile Laveleye and Léonce de Lavergne, *Études d'économie rurale. La Néerlander, Précédé du rapport de M. Léonce de Lavergne sur l'économie rurale de la Belgique* (Paris: A. Lacroix, Verboeckhoven et Cie, 1865). Swedish translations: *Nederländernas åkerbruk* (Lund, 1866) and *Åkerbruket i Belgien* (1867); Léonce de Lavergne, *Essai sur l'économie rurale de l'Angleterre, de l'Écosse et de l'Irlande* (1854). Swedish translation: *Skildring af landtmannaförhållandena inom Storbritannien och Irland jemförda med dem uti Frankrike* (Stockholm, 1856).

3 Angus Maddison, *Statistics on World Population, GDP and Per Capita GDP, 1-2008 AD*, http://www.ggdc.net/MADDISON/oriindex.htm.

4 Seppo Muromaa, *Suurten kuolovuosien* (1696-97) väestönmenetys Suomessa, Historiallisia Tutkimuksia 161 (Helsinki: SHS, 1991); Oiva Turpeinen, *Nälkä vai tauti tappoi? Kauhun vuodet, 1866-1868* (Helsinki: Societas Historica Finlandiæ, 1986); Marie Nelson, *Bitter Bread: The Famine in Norrbotten, 1867–1868* (Stockholm: Almqvist and Wiksell, 1988).

5 Daniel R. Headrick, *The Tentacles of Progress: Technology Transfer in the Age of Imperialism, 1850-1940* (New York and Oxford: Oxford University Press, 1988), 9-10.

6 George Basalla, *The Evolution of Technology* (Cambridge: Cambridge University Press, 1988); Robert A. Friedel, *Culture of Improvement: Technology and the Western Millennium* (Cambridge, MA: MIT Press, 2010).

7 Hughes, *Human-Built World*, 4.

8 Thomas J. Misa, *Leonardo to the Internet: Technology and Culture from the Renaissance to the Present* (Baltimore: Johns Hopkins University Press, 2004), 263.

9 Charles T. Stewart and Yasumitsu Nihei, *Technology Transfer and Human Factors* (Lexington, MA: Lexington Books, 1987), 1.

10 William H. Gruber and Donald D. Marquis, *Factors in the Transfer of Technology* (Cambridge, MA: MIT Press, 1969), 255.

11 Timo Myllyntaus, Minna Hares, and Jan Kunnas, "Sustainability in Danger? Slash-and-Burn Cultivation in Nineteenth-Century Finland and Twentieth-Century Southeast Asia," *Environmental History* 7, 2 (2002): 267-302.

12 Arvo M. Soininen, *Vanha maataloutemme, Maatalous ja maatalousväestö Suomessa perinnäisen maatalouden loppukaudella 1720-luvulta 1870-luvulle* (Helsinki: Suomen historiallinen seura, 1974), 54, 58, and 62-65; Arvo M. Soininen, *Pohjois-Savon asuttaminen keski- ja uuden ajan vaihteessa* (Helsinki: Historiallisia tutkimuksia, Suomen historiallinen seura, 1981), 137-62 and 256-58; Jouni Aarnio, *Kaskiviljelystä metsätöihin, Tutkimus Pielisjärven kruununmetsistä ja kruununmetsätorppareista vuoteen 1910*, Julkaisuja 4 (Joensuu, Finland: Joensuun yliopiston maantieteen laitos, 1999), 75-76; Matti Sarmela, "Swidden Cultivation in Finland as a Cultural System," *Suomen antropologi* 12, 4 (1987): 241-49.

13 Soininen, *Vanha maataloutemme*, 54, 58, and 62-65, and *Pohjois-Savon asuttaminen* 137-62 and 256-58; Aarnio, *Kaskiviljelystä metsätöihin*, 75-76; Sarmela, "Swidden Cultivation," 241-49.

14 Kari Tarkiainen, *Finnarnas historia i Sverige 1* (Helsinki: SHS, 1990), 148; Kjell Lööw, *Svedjefinnar: om 1600-talets finska invandring i Gävleborgs län* (Gävle, Sweden: Länsmuseet i Gävleborgs Län, 1985), 41. Tarkiainen's notion on the role of Duke Charles is supported by Eija Lähteenmäki, *Ruotsin suomalaismetsien synty ja savolainen liikkuvuus vanhemmalla Vaasa-kaudella* (Helsinki: Suomalaisen Kirjallisuuden Seura, 2002), 149, and questioned by Richard Broberg, *Finsk invandring till mellersta Sverige.* Svenska landsmål och svenskt folkliv. B 68 (Uppsala: Dialekt-och folkminnesarkivet, 1988).

15 Lööw, *Svedjefinnar*, 58-59; Juha Pentikäinen, "The Forest Finns as Transmitters of Finnish Culture from Savo Via Central Scandinavia to Delaware," in *New Sweden in America*, ed. Carol E. Hoffecker, Richard Waldron, Lorraine E. Williams, and Barbara E. Bens (Newark: University of Delaware Press, 1995), 291-301.

16 Tarkiainen, *Finnarnas historia i Sverige 1*, 148; Kjell Lööw, *Svedjefinnar*, 41.

17 E.A. Louhi, *The Delaware Finns or The First Permanent Settlements in Pennsylvania, Delaware, West New Jersey, and Eastern Part of Maryland* (New York, Humanity Press Publishers, 1925).

18 Asbjørn Fossen, *Historien om de norske skogfinnene* (Oslo: Asbjørn Fossen, 1992).

19 Gunlög Fur, *Colonialism in the Margins – Cultural Encounters in New Sweden and Lapland* (Leiden and Boston: Brill, 2006), 94-95; Per M. Tvengsberg, "Finns in Seventeenth-Century Sweden and Their Contributions to the New Sweden Colony," in Hoffecker et al., *New Sweden in America*, 279-90.

20 "Transumt av Kungl: Maj:ts Resolution, angående Finnarnes skadelige skogshyggen och svedjefällande; Dat den 22 Juni 1641," reproduced in Anders Björnson and Lars Magnusson, *Jordpäron – Svensk ekonomihistorisk läsebok* (Stockholm: Atlantis, 2011), 173.

21 Terry G. Jordan and Matti Kaups, *The American Backwoods Frontier – An Ethnic and Ecological Interpretation* (Baltimore: John Hopkins University Press, 1989); Tvengsberg, "Finns in Seventeenth-Century Sweden"; Fur, *Colonialism in the Margins.*

22 Jordan and Kaups, *The American Backwoods Frontier*, 8, 53-59, and 94; Terry G. Jordan, "New Sweden's Role on the American Frontier: A Study in Cultural Preadaptation," *Geografiska Annaler*, Series B, *Human Geography* 71, 2 (1989): 71-83; Terry G. Jordan, "The Material Cultural Legacy of New Sweden," in Hoffecker et al., *New Sweden in America*, 302-18; Frederick Jackson Turner, "The Significance of the Frontier in American History," in *The Frontier in American History* (Holt, Rinehart and Winston, 1962), 3; see also John Solomon Otto, "The Diffusion of Upland South Folk Culture, 1790-1840," *Southeastern Geographer* 22, 2 (1982): 89-98.

23 Nils-Arvid Bringéus, *Brännodling – En historisk-etnologisk undersökning* (Lund, Sweden: CWK Gleerup, 1963), 149.

24 Stephen J. Pyne, *Vestal Fire – An Environmental History, Told through Fire, of Europe and Europe's Encounter with the World* (Seattle: University of Washington Press, 1997), 169-70.

25 Bringéus, *Brännodling*; Jacob Serenius, *Engelska åkermannen och fåraherden* (Stockholm, 1727); Stephen Bennet, *Sten Biörnsons jordmärg, eller trogna bondaläro* (Stockholm, 1745).

26 Thaer, *Den rationela landthushållningens grundsatser.*

27 Ulrich Lange, *Experimentalfältet: The Royal Swedish Academy of Agriculture's Experimental Station in Stockholm, 1816-1907* (Uppsala: Kungl. skogs-och lantbruksakademien, 2000).

28 Pehr Adrian Gadd, *Inledning till Svenska Landt-Skötseln*, part 1 (Stockholm, 1773), 260-68; Jacob Tengström, *Försök till lärobok i landthushållningen, för finska bonden* (Åbo, 1803).

29 K.R. Melander, "Några uppgifter om äldre tiders kärrodling i vårt land," *Historisk tidskrift för Finland* 7 (1922): 24; Armas Lukko, *Etelä-Pohjanmaan Historia III: Nuijasodasta Isoonvihaan* (Vaasa, Finland: Etelä-Pohjanmaan Historiatoimikunta, 1945), 99-104.

30 Urban Hiärne, *Den besvarade och förklarade anledningens andra flock om jorden och landskap i gemeen* (Stockholm, 1706).

31 Lukko, *Etelä-Pohjanmaan Historia III*, 99-104.

32 Soininen, *Vanha maataloutemme*, 145 and 148; A.R. Saarenseppä, *Kuvauksia Pohjois-Karjalan maataloudellisista oloista vuoden 1800:n vaiheille* (Joensuu: Pohjois-Karjalan maanviljelysseura, 1912), 131.

33 Helmer Smeds, *Malaxbygden* (Helsinki: University of Helsinki, 1935), 131-32.

34 Mikael Wexionius, *Epitome descriptionis Sveciæ, Gothiæ, Fenningiæ et subjectarum provinciarum* (Aboæ, 1650). Translation from Latin by Gösta Grotenfelt in *Det primitiva jordbrukets metoder i Finland under den historiska tiden* (Helsinki, 1899), 28.

35 Grotenfelt, *Det primitiva jordbrukets metoder i Finland*, 205.

36 Soininen, *Vanha maataloutemme*, 52; Kustaa Vilkuna, *Isien työ, Veden ja maan viljaa, arkityön kauneutta* (Helsinki: Otava, 1976), 167.

37 Grotenfelt, *Det primitiva jordbrukets metoder i Finland*, 205-10; Soininen, *Vanha maataloutemme*, 52.

38 P.A. Tamm in *Upsala läns kgl. landthushållnings-sällskaps handlingar för 1819* (Upsala: Bruzelius, 1820), 15-20.

39 E. Bergström in *Upsala läns kgl. landthushållnings-sällskaps handlingar för 1820* (Upsala: Palmblads, 1821), 23-24.

40 Nils-Arvid Bringéus, "The Paring and Burning Spade in Sweden," in *The Spade in Northern and Atlantic Europe*, ed. Gailey and Fenton (Belfast: Ulster Folk Museum, Queen's University of Belfast, 1970), 88-98; Erik Bylund, *Koloniseringen av Pite Lappmark t.o.m. år 1867* (Uppsala: Almqvist and Wiksells Boktryckeri, 1956); Jan Kunnas, "A Dense and Sickly Mist from Thousands of Bog Fires," *Environment and History* 11, 4 (2005): 431-46.

41 "Beskrifning öfver Löfångers socken i Vesterbottens Län inlemmad till Länets Kongl: HushållsSällskap, år 1827," in *Vesterbottens Läns Kongliga Hushålls-Sällskaps Handlingar*, vol. 2 (Stockholm: Zacharias Hæggström, 1829), 71-72 (trans. by the author); see also *Vesterbottens Läns Kongliga Hushålls-Sällskaps Handlingar*, vol. 1 (Stockholm, 1827), 89.

42 *Vesterbottens läns kongliga hushålls-sällskaps handlingar*, 2: 71.

43 Magnus Stridzberg, *En grundelig kunskap om swenska åkerbruket* (Stockholm, 1727), 56-57.

44 Johan Petter Arrhenius, "Om hvitmossorna, deras nytta och tillgodogörelse för vårt lantbruk (jemte anvisning om fördelaktigs sätt för användande af bättre mossar och kärrtrakter, som ej kunna afdiktas)," in *Smärre Samlade Skrifter i Landthushållningen*, vol. 4 (Stockholm, 1865) (transl. by the author); see also Pehr Adrian Gadd, *Inledning till Svenska Landt-Skötseln*, part 1, (Stockholm, 1773), 250-51.

45 Kunnas, "A Dense and Sickly Mist"; Jan Kunnas and Timo Myllyntaus, "Postponed Leap in Carbon Dioxide Emissions: Impacts of Energy Efficiency, Fuel Choices and Industrial Structure on the Finnish Energy Economy, 1800-2005," *Global Environment* 3 (2009): 154-89.

46 Compare with David Edgerton, *The Shock of the Old: Technology and Global History since 1900* (New York and Oxford: Oxford University Press, 2006).

47 Kunnas, "A Dense and Sickly Mist"; Maisema-aluetyöryhmä, *Maisema-aluetyöryhmän mietintö osa I, Mietintö 66/1992* (Helsinki: Ympäristöministeriö, Ympäristönsuojeluosasto, 1993), 13; Pertti Rassi, Esko Hyvärinen, Aino Juslén, and Ilpo Mannerkoski, eds., *The 2010 Red List of Finnish Species* (Helsinki: Ympäristöministeriö and Suomen ympäristökeskus, 2010).

48 Torsby Finnkulturcentrum, http://www.varmlandsmuseum.se/ and Finnskogsmuseet in Skräddrabo, http://home.swipnet.se/finnskogsmuseet/.

The Sheep, the Market, and the Soil
Environmental Destruction in the Icelandic Highlands, 1880-1910

ANNA GUDRÚN THÓRHALLSDÓTTIR, ÁRNI DANÍEL JÚLÍUSSON, and HELGA ÖGMUNDARDÓTTIR

The Icelandic highlands form an especially sensitive ecosystem. The consensus has been that erosion because of livestock grazing began as early as the tenth century, and that this was the main cause of erosion.[1] However, the present consensus about the effect of livestock grazing on Icelandic vegetation before 1800 is in need of a review. The impact of volcanic activity is increasingly understood as the main culprit of erosion before the nineteenth century.[2] Recent research also suggests that there were long periods of regeneration, for example, in the fifteenth through seventeenth centuries.[3] As well, the impact of a cooler climate in the late thirteenth century seems to have been underestimated.[4] On the whole, the size of the livestock population before the nineteenth century was not large enough to greatly impact the soil cover in the highlands, even if some erosion certainly occurred.[5] In 2001, I.A. Simpson and colleagues wrote: "There remains therefore a dilemma in how to explain the relationship between soil erosion and sheep grazing in South Iceland when there is strong evidence for sufficient biomass to support the numbers of livestock. The issue of soil erosion and its relation to livestock is clearly of greater subtlety than previously considered."[6] In the nineteenth century, as economic growth began in Iceland, society began to change.

From the 1820s, foreign markets opened up for sheep products like wool and tallow, especially in Denmark.[7] And the number of sheep began to rise. From the 1880s, lucrative markets opened up in Great Britain for live sheep.[8]

As the number of sheep increased further, the soil cover began to suffer at the same time as the climate turned colder. In 1897, the market for sheep in Great Britain suddenly disappeared, but new markets were found for sheep products like wool and meat, especially for salted mutton in Denmark and Norway. Along with a burgeoning economy, the population in Iceland grew and moved to towns, these towns also becoming markets for mutton.

This chapter highlights the main nuances of climatic changes and changes in land use between 1880 and 1910, when soil erosion truly became a devastating man-made force of nature. The central question we ask is, is it possible that the opening of Iceland as a commodity frontier in sheep products in the world economy was the main culprit in ecodestruction in Iceland in the nineteenth and twentieth centuries? Three factors are considered in turn: the environmental factors causing the soil erosion, the traditional farming techniques in Iceland, and the growing trade in sheep products. The different factors causing erosion are then compared and weighed against each other.

The Environment of Iceland

Volcanic Activity

Iceland lies on the junction of the Mid-Atlantic Ridge and the Greenland-Iceland-Faroes Ridge. The interaction between the spreading plate boundaries and mantle plume results in active volcanism and gives the island its volcanic character.[9] Volcanism in Iceland features nearly all volcano types and eruption styles known on earth. Icelandic eruption history during historical times is well documented not only in volcanological records but also in written documents extending back to the eleventh century. A comprehensive analysis of these in 2007 lists 205 volcanic events in historical times; since then, three eruptions have taken place, including Eyjafjallajökull in 2010.[10] The volcanic activity has, however, been variable from one time to another, being very low in the first centuries after settlement and the centuries thereafter, and again in the sixteenth century (Figure 7.1). The activity seen as number of eruptions is not a measure of the scope of the destruction caused by volcanic activity, as the consequences of the eruptions are also variable. In historical times, the most destructive eruptions happened in the twelfth century (Hekla in 1104), fourteenth century (six major eruptions, including Öræfajökull in 1362), and eighteenth century (Mývatnseldar in 1724-29 and Skaftáreldar in 1783-84).

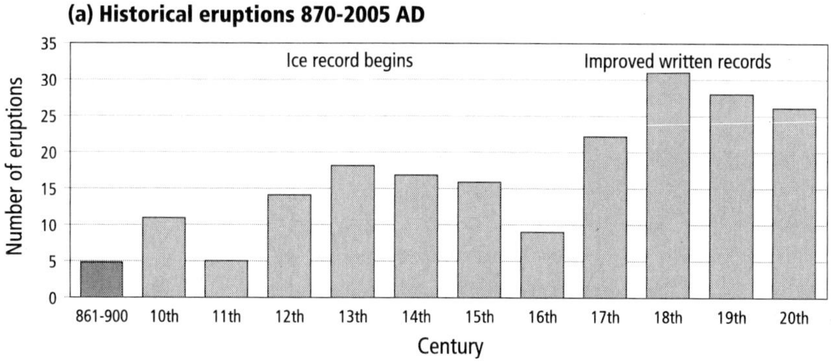

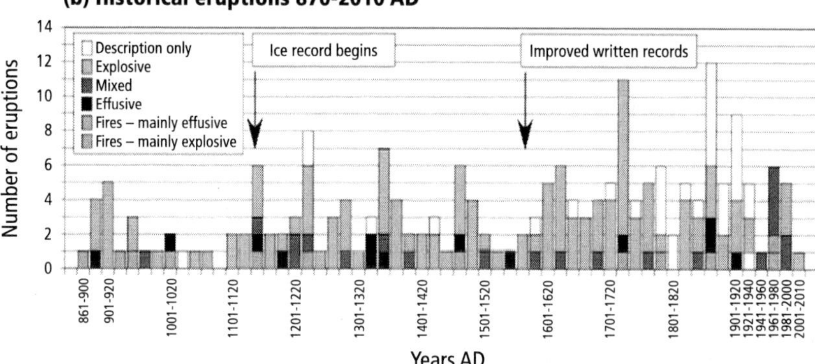

Figure 7.1 Known volcanic eruptions in Iceland throughout history. Based on T. Thordarson and G. Larsen, "Volcanism in Iceland in Historical Time: Volcano Types, Eruption Styles and Eruptive History," *J. Geodynamics* 43 (2007): 118-52.

Soil Erosion

The volcanic origin of Iceland is manifested in the dominating andosols soil types, with high erosion susceptibility.[11] In most eruptions, airborne tephra is carried, often long distances, and will deposit in sinks in the landscape. This drifting tephra causes destruction not only while moving but also where it accumulates, by choking underlying vegetation, which easily leads to soil erosion. Soil erosion can be estimated by measuring the rate of soil thickening between known tephra layers (tephrachronology) or from ocean and lake sediment thickening and composition.[12] It has generally been accepted that the main cause of land degradation and soil erosion in Iceland was livestock grazing.[13] More recent sediment data reveal, however, that soil

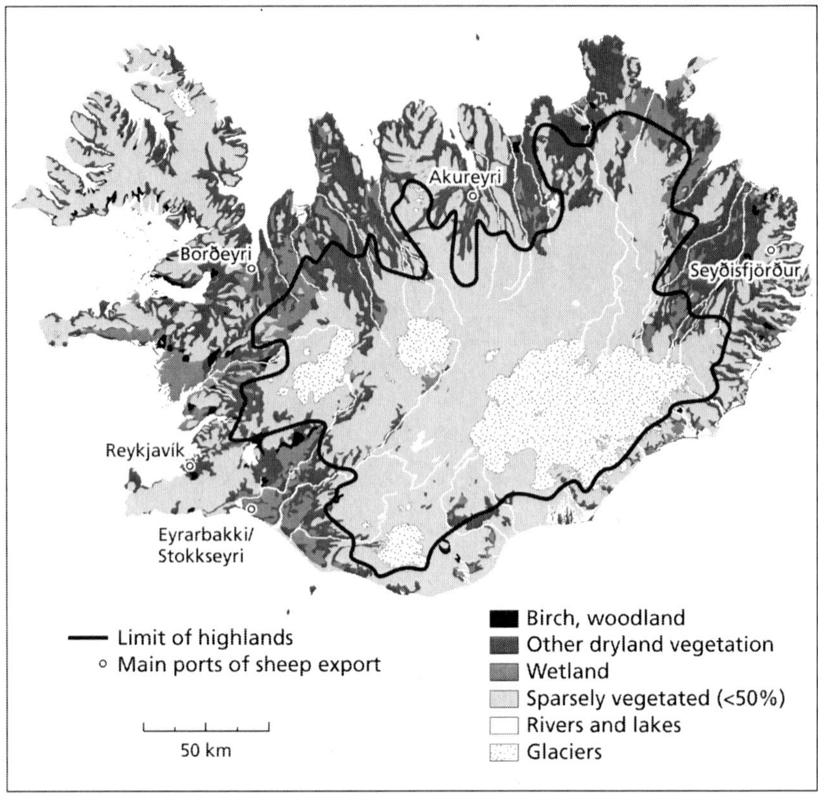

Map 7.1 Iceland's vegetation zones, showing the delimitation of the central highlands. Source: Icelandic Institute of Natural History, 2001.

erosion in Iceland existed and occurred often before settlement and correlates with major eruptions with tephra fallouts and climate cooling.[14] Just before or by the time of settlement (c. AD 871), a major eruption occurred. The tephra layer from that eruption is found over most of Iceland. In some catchments it forms the thickest tephra layer in the soil profile from the last two thousand years.[15] The effect of this fallout, just before settlement, has probably been underestimated, and the presumed initiation of erosion in Iceland by settlement confounded with this tephra fallout. In support for this interpretation, there is no evidence for increased soil erosion after AD 900, and from the data it seems clear that the rate of soil thickness, both in south and north Iceland, was fairly constant after settlement, but increased after the major eruption in Hekla in 1104.[16] Data indicate that greatest soil erosion (measured as terrestrial organic carbon flux to southwest Iceland ocean shelf) occurred from about 1780 to about 1920.[17]

Climate

The climate from the time of settlement to present has been constructed from numerous documentary data and several measured environmental proxies.[18] It is clear that the climate in Iceland and the North Atlantic got abruptly cooler around 1250, at the onset of the Little Ice Age (LIA), which lasted until the first part of the twentieth century.[19] During this period, less-severe summer cold seems to have existed in the sixteenth and seventeenth centuries, followed by cooling again in the last part of the eighteenth century, with especially cold periods around 1840 and 1890.[20] The LIA was likely the coldest time during the last 8,000 years in Iceland, with average temperatures of 1 to 2°C, which can be compared to the 1961-90 average of 3.6°C.[21]

Recordings of meteorological observations in Iceland date back to 1798 in Stykkishólmur, west Iceland.[22] These recordings show that mean annual temperatures were very low in the latter part of the nineteenth century, with two years below 1°C mean annual temperature (1859 and 1866) and several years not reaching 2°C. In comparison, in 2000 and 2010, the mean annual temperature in west Iceland in 2003 was 5.4°C and 4.7°C.

Vegetation

Before the time of settlement there were no ungulate grazers in Iceland. The vegetation was fragile, lacked a defence mechanism, and was dominated by grazing-sensitive species (Map 7.1).[23] The Book of Settlements, *Landnámabók*, states that when the Norwegians came to Iceland in 872, the country was "covered with wood between fell and foreshore."[24] The surviving versions of *Landnámabók* date from the second half of the thirteenth century, but the book is believed to have been composed by Ari Þorgilsson, who lived from 1068 to 1148. As it is assumed to have been written more than two hundred years after settlement, it cannot be considered a reliable source,[25] even though it is most often referred to as the main reference for pre-settlement vegetation in Iceland, giving the best reflection on "natural" or "climax" vegetation composition.[26] Pollen records from several sites in Iceland reveal that birch (*Betula pubescens*), for instance, had a much more widespread distribution in Iceland before settlement.[27] However, the same records also indicate a retrogressive succession of the birch woodland decades or even centuries before settlement and a quite gradual contraction, rather than a rapid change, of the woodlands over the course of several hundred years after settlement.[28] Both pollen diagrams and macrofossil evidence reveal an increase in grasses and sedges along with a decline of woodland species following settlement and, because of development of more varied habitats, increased local plant diversity.[29] Both grasses and sedges are, in general, grazing-tolerant taxa, and this change reflects anthropogenic influences and an adaptation of the Icelandic ecosystem to land-use changes. Natural grasslands and meadows are productive ecosystems – with a high rate of biogeochemical cycling, transpiration, and biodiversity – that are created and maintained by grazing.[30]

Winter Fodder Production

According to Páll Bergþórsson, every one degree Celsius lower or higher decreases or increases, respectively, primary production and therefore the carrying capacity of the land by at least 15 percent. Further, as the temperature falls, the nutrient requirements of livestock increase, and during low winter temperatures there can be a 10 to 15 percent increase in fodder consumption.[31] As the climate got colder in Iceland after 1250, cultivation became difficult, and winter fodder was increasingly collected from outfield wetlands and meadows. This collection was very limited, and supplementary feed during winter was primarily for dairy cattle, while the sheep mostly grazed during winter, with limited supplementation. Because of this higher cost of holding dairy cattle, with time, the sheep took over as a milk-producing animal. Horses did not generally receive supplementary fodder – only a few riding horses in use did. In harsh winters, the flock would collapse, and from written resources it has been calculated that in the very cold period in the nineteenth century the total livestock flock would diminish by 30 percent for every one degree Celsius lower annual temperature.[32] Thus, throughout the centuries, the grazing pressure on the land was regulated by the number of animals that would survive the winter: in cold periods, the grazing pressure would be lower and therefore trail the carrying capacity of the rangeland. This relationship broke, however, with initiation of pasture cultivation in the late nineteenth and early twentieth centuries, as setting off winter fodder (hay) made it possible to keep higher numbers of livestock throughout the year.

Traditional Land Use in Iceland

Agricultural Technology

Agricultural technology in Iceland was based on the infield-outfield system, developed in northwestern Europe, including western Norway, in the period AD 200-800, with a basis in earlier agricultural systems.[33] Its basic characteristics were a mixed economy with husbandry and the growing of grains. Cattle were kept in byres during the winter, and cattle dung was collected to fertilize the infield, a continually used field for grain growing. The cattle were milked in the summer but usually not in the winter. Some sheep were kept, probably mostly for the wool, but some were also milked. The proportional number of heads of cattle to sheep was about one to one, with the result that most of the production was from the cattle, because one cattle produced about the equivalent of six sheep. Most of the vegetation resources of each farm were in the outfield, where the livestock was grazed and hay for the winter collected. This kind of agricultural technology existed in Norway and on the British Isles of the Orkney and Hebrides around AD 900, when it was imported to Iceland and continued developing there. Among the more important modifications to the system in Iceland was the increase in number of sheep proportional to cattle. This happened primarily after 1400, when repeated pandemics led to labour shortages. Cattle were less productive than sheep per input of working hours, so in a situation of labour shortage, the number of sheep grew and the number of cattle declined.[34]

Grazing Land

From soon after the time of settlement in the ninth century, the use of land for grazing was highly regulated. In *Grágás* and *Jónsbók,* Icelandic law books from the twelfth century, there are strict provisions for grazing management; these provisions were effective in Iceland until 1969, when new grazing acts came in force.[35] In many ways, the old grazing provisions are much more accurate and strict than the new ones and regulated more precisely the grazing pressure on the land.

Grazing land is and has always been a very important but limited natural resource. Already in *Jónsbók* (1281) there is a set payment for the grazing resource: one lamb for each sixty lambs on the land.[36] By the mid-eighteenth century, the payment had risen to one lamb for every twenty. With this economical interest, churches had gained ownership of extensive grazing areas, both in the highlands and the lowlands, and collected the payment from these. Thus, in the latter part of the nineteenth century when the interest of the farming community in the grazing land increased, the Church controlled much of the grazing land. From 1850, and especially between 1880 and 1900, farming communities all over Iceland bought grazing land, mainly from churches but also from prosperous farms that owned good grazing areas in the highlands. These grazing areas became a common grazing area for the farms in each community. Many of these areas were in far-off highlands that had not been under regulated grazing earlier.[37]

The number of cows at each farm was usually about two to six, depending on the size and location of the farm, and the area of the country. The number of ewes was twenty to sixty, also depending on the farm's size and location. The number of horses was very different from county to county, the largest number being in the counties of Árnessýsla and Rangárvallasýsla, in the south lowlands, and Húnavatnssýsla and Skagafjarðarsýsla, in the northwest.[38]

In the eighteenth century, the typical Icelandic sheep flock consisted of lactating ewes, dry ewes, yearling replacements, and wethers (castrated males). Lactating ewes made up more than 50 percent of the flock, wethers about 20 percent.[39] Wethers were kept for wool and meat production, whereas the ewes were kept for the milk. In *Jónsbók,* there is a provision stating that "only one tenth of ewes are allowed to go with their lambs" – or in other words, 90 percent of the ewes should be kept home (or in shielings) and milked.[40] Although unpopular, this provision was adhered to more or less throughout the centuries, and the Icelandic sheep must be classified as a dairy breed up to the twentieth century.

Lambs were born in May and kept with their mothers for five to eight weeks, until sometime between 24 June (Jónsmessa – Midsummer) and 2 July (Þingmaríumessa).[41] The ewes were kept by the farm or on a summer farm (*sel*) for milking. As the ewes were milked twice a day (for nine to thirteen weeks), they had to be kept close to the farm, and usually a child or a youngster would go with the flock and direct it out to the grazing areas, which would be in the vicinity of the farm. Thus, the main sheep flock, i.e., lactating ewes, did not roam freely for grazing but were under regulated grazing in the lowlands. The remaining sheep flock consisting of wethers, dry ewes, yearlings, and lambs were driven on rangeland pastures, in mountains, and in highlands for grazing. The grazing pressure from this part of the sheep flock was comparatively small in relation to the grazing pressure of the whole flock, as the producing ewes have the most grazing intake, dry ewes and wethers less, and lambs very little their first summer. Thus, the grazing pressure on the mountain and highland areas in Iceland was low at the beginning of the eighteenth century compared with what it became at the end of the nineteenth century.

Consumption and Trade

It is estimated that in the mid-eighteenth century over half of Icelandic food consumption was of milk and milk products.[42] On many farms, especially smaller and poorer farms, sheep milk, butter, *skyr* (fresh yogurt), and whey were the most important products on the farm, both for domestic consumption and to pay dues for the farm. Although cows produced a much higher quantity of milk – a cow could be expected to produce one to two thousand litres of milk per year, whereas a ewe produced about thirty to forty litres per year – the milk of the ewes contained twice as much milk fat per litre as cow's milk. The ewe's milk was also a great component of the overall usefulness of sheep, along with their wool, tallow, and meat. Before the opening of the commodity frontier, trade was minimal and mostly organized around the needs of Icelandic landowners, who sold to foreign traders fish and products the peasantry had paid in land rent. Some of the peasants could also sometimes barter a little of their own products for grain and other non-local commodities.

Commodity Frontiers

The world commodity market came into existence in the sixteenth century with the production of a wide range of products in several parts of the world and organized by the European powers for profit. Sugar was produced in the West Indies with slave labour, timber came from Norway to the Dutch Republic to build ships to sail slaves bought in Africa to the West Indies.[43] By 1800, the Industrial Revolution was gaining speed. It had begun in Great Britain in the eighteenth century with the increased use of coal and steam machines in producing goods. After 1800, feudalism disappeared from continental Europe, and trade, industrial growth, and economic development in general gathered pace, slowly at first but gaining momentum after 1840.

The environment historian Jason W. Moore has set out to restore the concept of frontier for the study of world capitalist expansion.[44] His goal is to suggest ways of rethinking early modern capitalist expansion as a socio-ecological process. He reconceptualizes the term "frontier" within the world-systems paradigm, a paradigm developed primarily by Immanuel Wallerstein on the basis of the Marxist analysis of capitalism.[45] Moore points out that the conceptualization of frontier as the area where incorporation into the capitalist system occurs is insufficient because it fails, for example, to distinguish the incorporation of the Americas from Asia and Africa, where strong state structures impeded full incorporation until the nineteenth and twentieth centuries. Moore wants to draw attention to the ways in which "the production and distribution of specific commodities, and of primary products in particular, have restructured geographic space at the margin of the system in such a way as to require further expansion."[46] He suggests a reconceptualization of the concept of frontier by adding the word "commodity," thus the commodity frontier. In commodity frontiers, ecological exhaustion was not only a fact of life but also a major impetus to further capitalist expansion: "Ecological exhaustion at the point of production was complemented by an environmentally destructive multiplier effect which led to, inter alia, deforestation, massive soil erosion, siltation, climate change, and other effects in the case of sugar."[47]

Multiple commodity frontiers existed in the Americas: sugar, silver, timber, cattle, foodstuffs, cotton, and so on. Harold Innis developed a theory similar to the commodity frontiers concept called the staple approach, in which he analyzed the economic development of his subject areas through the development of staples like fur or cod. In the case of fur, he famously suggested that the fur trade largely determined Canada's boundaries.[48]

The idea of mining vegetable resources has been a basic paradigm in environmental historical narratives of Iceland.[49] According to this narrative, Icelandic resources were quickly devoured by the settlers of the Viking Age, leading to an environmental catastrophe. However, based on Moore's work, it is doubtful if an early modern ecologically destructive commodity frontier should be conceptually transferred to the ninth and tenth century, with its early feudal economy. Moore suggests that the contemporary global ecological crisis is not rooted in the Industrial Revolution per se but in the logic of capital itself. With the creation of the world market and a trans-Atlantic division of labour in the sixteenth century, the products of the countryside flowed into the cities. Nutrients were pumped out of one periphery and transferred to the core. The land was essentially progressively mined, until its relative exhaustion fettered profitability. With Moore's idea of commodity frontiers in mind, the staple of sheep products like wool, tallow, or even live sheep can be analyzed in a new way: the ecological dimension of a transnational commodity chain involving sheep products.

The Frontier Opens

The development of commodity frontiers in Iceland occurred under the control of a strong state. This state, the Danish-Norwegian kingdom, had until about 1800 controlled and fettered all attempts at establishing capitalist enterprise in the Icelandic context. Then, and especially after 1855, the state felt confident enough to give free rein to capitalist development in Iceland. This was supported by the Icelandic nationalist movement; indeed, its main spokesman, Jón Sigurðsson, wrote extremely positively about the value of economic development based on the linkage to English economic development.[50] The development of the commodity frontier of sheep products was a key moment in the political and economic development of Iceland. However, in the long run, it led to environmental problems.

The period from 1820 to 1860 was a period of expansion in Iceland. The number of inhabitants rose, as did the number of animals and farms. It was also a period of qualitative change in the connection of the Icelandic farmers to the world market. After 1820, the number of ships and tonnage importing grain to Iceland rose steadily.[51] In Iceland, these ships were loaded up with a variety of products, most of them new commodities Iceland had never produced before in any great quantity: salt fish from the southwest, wool and tallow from every part of the country, and shark liver oil from the Westfjords and northern Iceland.

The origin and causes of this expansion of trade, which was directed by Danish merchants operating from Copenhagen, has not been the focus of any research, nor has its effect on the environment. A comparison of the trade of ordinary farmers in 1760 and those in 1840 shows a great difference. In 1760, half to two-thirds of all farmers participated in no trade at all. The amount traded by farmers who did was very small, sometimes as little as half a barrel of rye for some few wethers or a little fish. In 1840, on the other hand, every farm would trade large amounts of wool and tallow, and receive several barrels of rye, wheat, and beans. Coffee, sugar, and tobacco also had become consumption items of Icelandic farmers and their families.[52] On the whole, the main sheep product for export was wool, and with increasing numbers of sheep, wool export increased, from 130,152 kilograms in 1806 to 582,012 kilograms in 1862.[53]

The expansion of the sheep commodity trade led to the expansion of sheep farming. The number of sheep increased through the nineteenth century: in 1800, there were 304,000; in 1826, 402,000; in 1840, 505,000; and in 1896, 589,000.[54] The number of ewes in Iceland in 1800 was about 170,000, but by 1850 it had grown to about 260,000.[55] The number of sheep per inhabitant rose in the same period from about five to more than eight. This meant that the average Icelander had become richer. A surplus had been created, and Icelanders were now participating in the great world economic expansion of the nineteenth century.

In 1861, a booklet was published that encouraged farmers to increase sheep numbers as much as the land allowed because "if there was enough to sell[,] the English would buy."[56] When wool prices collapsed in Britain in

1867 other markets for Icelandic sheep products needed to be found. In the 1870s, British merchants showed up in Iceland and started buying sheep that they transported live on ships to England and Scotland. In this way they could slaughter the sheep when there was demand and offer the meat fresh instead of salted. In 1876, 3,100 live sheep were exported. That number skyrocketed to 82,818 in 1890, when nearly 25 percent of export value from Iceland was from live sheep.[57] The trade was even more important for Iceland and the farming community than these numbers indicate. The British sheep trading was not barter exchange, like all trade had been in Iceland for centuries. Rather, the sheep were often paid for in gold and silver coins, these coins being the first real currency that the Icelandic farming community experienced. It has been argued that this capital was the basis for the founding of the first Icelandic bank in 1886.[58] The Danish merchants, who had a de facto monopoly on the Iceland trade until 1855, did not pay in cash, only in products. The Danish trade was therefore a barter trade: the farmers had to accept goods like rye or coffee for their sheep products, rather than gold or silver coins. They also had to take their sheep flocks to the Danish merchants and slaughter the sheep themselves, unlike with the British merchants, who came to the farmers' home area to buy the live sheep flocks, which was much more convenient for the farmers.[59]

Technological Change

The case of the foreign trade in sheep in Iceland in the nineteenth century is an interesting case of demand preceding technological change. The capitalist world market (in the Icelandic context) first utilized the space for growth by co-opting existing technologies, be it the existing agricultural technology or the existing communication technology of sailing ships for commerce, and then new technologies were used to facilitate further growth.

In about 1800, the proportion of cattle to sheep was about one to ten. The basic logic of the agricultural system was still the feudal one of keeping enough animals to support the family and pay the taxes and land rent to the landowner. This meant that the size of the herd was determined by the size of the family and the size of the tax and land rent, rather than by how much it was possible to produce on the existing grazing. It was this that changed after 1800. The herd was now expanded in response to world market demand. The possibilities opened by the market led to yet another change in the proportion of cattle to sheep, now sometimes as high as one to twenty. The sheep, or some of the sheep, were the commercial part of the herd, but the cattle did not enter that sphere. The flock of sheep could be expanded because of unused space in the highlands.

After 1880, further expansion of the production of sheep products for the world market led to much more radical changes in technology. The expansion of cultivated hay meadows with plowing and harrowing began; barbed wire was introduced; and after 1910, so too was artificial fertilizer. Artificial fertilizer was a revolutionary product, releasing agriculture from hindered development of agricultural productivity caused by thousands of years of fertilizer shortage.

An analogous development occurred in communications technology. At the beginning of the nineteenth century, sailing ships imported goods to Iceland and exported fish, sheep products, and other Icelandic articles. After 1820, the tonnage of sailing ships serving Iceland suddenly increased.[60] Many more and bigger ships now came to Iceland each year, and the export capacity of those ships expanded greatly. These were all sailing ships, which were subject to whatever wind conditions prevailed, and this likely fettered the expansion of trade. Steam ships were introduced between 1860 and 1870, and it might be thanks to them that it became possible to export live sheep to Great Britain, as these ships could travel with much more regularity and security than the sailing ships.

Expansion of Grazing in the Highlands

Change of Traditional Land Use

Since the ewes were kept for milk production, they had to be kept close to the farm during the milking period, which lasted throughout the grazing period. With the opening of the frontier and British merchants coming to Iceland, the role of sheep in the Icelandic farming system changed. The interest was no longer in sheep milk or milk products but in meat that could be sold for currency. Now, it was most important to get as much growth in the lambs for slaughter the first year, and for that good grazing areas were a necessity. These were found in the extensive highlands (Figure 7.2). An example of these changes in the use of grazing areas in the highlands in the second part of the nineteenth century is the case of Gnúpverjahreppur, a rural municipality in Árnessýsla, in southern Iceland, that has been described by Helga Ögmundardóttir.[61]

In Gnúpverjahreppur in the eighteenth and the first part of the nineteenth century, ewes were grazed during the summer close to the farm,

Figure 7.2 Sheep farming at the Hunkubakkar farm in Síða in southeastern Iceland. Movable sheep fences were used for fertilizing hay meadows. The photograph was taken by Eggert Guðmundsson, brother of farmer Jóhannes Guðmundsson (seated, on the right), in 1903. Originally reproduced in Daniel Bruun, *Íslenskt þjóðlíf í þúsund ár* (Reykjavik, Örn og Örlygur, 1987). Photograph in public domain.

while wethers and non-lactating ewes were driven to the highlands. These animals, however, never did go far into the highlands, and there was no need to search for sheep close to the glaciers; the sheep gathering in the fall did not go far inland. In about the mid-nineteenth century, sheep numbers had grown, and the sheep started to graze further into the highlands. In 1847, it was necessary for the first time to search for sheep further inland, north of the Dalsá River, during the fall sheep gathering. From 1850 to 1880, the number of sheep driven to the highlands increased. Fewer and fewer ewes were kept home by the farm for milking but were instead driven with their lambs to the highlands for grazing. After 1870, it became clear that the sheep went all the way to the wetlands by the glacier Hofsjökull, in the middle of Iceland. It became necessary to expand the organized sheep gathering and, in 1878, these wetland areas were searched for the first time. The area that had to be searched during the fall sheep gathering therefore gradually expanded during the second half of the nineteenth century.[62] The same change in land use was seen in other parts of the country in the latter part of the nineteenth century and the first decades of the twentieth century.

Soil Erosion

As mentioned earlier, the climate in the latter part of the nineteenth century was especially unfavourable in Iceland, with two years below a mean annual temperature of 1°C (1859 and 1866) and several years not reaching 2°C. This cold climate occurred when there was not only increasing numbers of sheep in Iceland but also a major change in land use – from mainly lowland summer grazing by the sheep to mainly inland/mountain summer grazing. In Iceland, inland/highland areas that were used for summer grazing in the late nineteenth century lie at around four to six hundred metres above sea level. It can be estimated that with every hundred metres of elevation, the temperature decreases by 1°C, leading to a shorter growing season and, accordingly, lower primary production. Therefore, the increased number of sheep in the inland/highland areas at the same time as the climate became especially unfavourable could be expected to have a disastrous effect. In the last part of the nineteenth century, erosion and drifting sand became an enormous problem, much more so than in the decades before, leading to the destruction of numerous farms in southern Iceland.[63] This environmental catastrophe led, in 1907, to the establishment of the Icelandic Soil Conservation Service, the first of its kind in Europe, to combat the erosion.

Conclusion

In Iceland, many factors have been blamed for the erosion that has led to the loss of vegetative cover: fragile soils, vulnerable vegetation, cold climate, and overgrazing. However, history shows that the unfortunate timing of increasing sheep numbers and simultaneous change of grazing practices – that is, increasing use of the highlands – that coincided with the extremely cold climate at the end of the nineteenth century led to an environmental disaster. Throughout the nineteenth century, the farming community in Iceland was undergoing changes, increasing the number of sheep, as well as changing farming practices. But it is likely that the new interest of British traders in buying sheep from Iceland for the growing industrial cities of Britain augmented the changes, which in turn increased the catastrophic result for the Icelandic highlands.

Was the opening of Iceland as a commodity frontier in sheep products for the world economy the main culprit in ecological destruction in Iceland in the nineteenth and twentieth centuries? This is the question we posed at the beginning of the chapter, and it can now be answered in the affirmative. The increase in the number of sheep in the nineteenth century, thanks to the growing demand for sheep products in the world market, led to a sudden increase in highland grazing, which had until then been very limited or non-existent. The period of very cold weather that hit Iceland starting in about 1880 led to a catastrophic destruction of vegetation in the highlands. The commodity frontier of the sheep in the Icelandic highlands had hit its limit. It might not be a coincidence that in about 1880 a new commodity frontier was opened in Iceland, this time in the sea. From 1880, cod fishing increased rapidly in the north and the east, where it had been limited before, and where the sheep trade expansion had been most pronounced.

In the nineteenth century, the Danish state in Iceland opened its borders to the development of commodity frontiers. This was enthusiastically supported by Icelandic nationalists, who saw great possibilities for economic gain. Indeed, the economic growth of the nineteenth century has been seen as the main prerequisite for the political independence of Iceland. The sheep trade with Britain led to increasing grazing in the highlands, and when the weather turned colder after 1880, these two factors – grazing and cold – caused an immense increase in erosion. For the first time, grazing had become a major cause of erosion in Iceland as a direct consequence of the opening up of the commodity frontier.

NOTES

1 Guttormur Sigbjarnarson, "Áfok og uppblástur: Þættir úr gróðursögu Haukadalsheiðar," *Náttúrufræðingurinn* 39, 2 (1969): 68-118; Rannveig Ólafsdóttir, "Land Degradation and Climate in Iceland: A Spatial and Temporal Assessment" (PhD diss., Lund University, 2001); Sigurður Þórarinsson, "Sambúð lands og lýðs í ellefu aldir," in *Saga Íslands*, vol. 1, ed. Sigurður Líndal (Reykjavik: Icelandic Literary Society and Historical Society, 1974), 29-97.

2 I.T. Lawson, F.J. Gathorne-Hardy, M.J. Church, A.J. Newton, K.J. Edwards, A.J. Dugmore, and A. Einarsson, "Environmental Impacts of the Norse Settlement: Palaoenvironmental Data from Mývatnssveit, Northern Iceland," *Boreas* 36 (2007): 1-19; A. Geirsdottir, G.H. Miller, T. Thordarson, and K.B. Olafsdottir, "A 2000 Year Record of Climate Variations Reconstructed from Haukadalsvatn, West Iceland," *Journal of Paleolimnology* 41 (2009): 95-115; D.J. Larsen, G.H. Miller, A. Geirsdottir, and T. Thordarson, "A 3000-Year Varved Record of Glacier Activity and Climate Change from the Proglacial Lake Hvítárvatn, Iceland," *Quaternary Science Reviews* 30 (2011): 2715-31.

3 Sigbjarnarson, "Áfok og uppblástur"; Sverrir Aðalsteinn Jónsson, "Vegetation History of Fljótsdalshérað during the Last 2000 Years: A Palynological Study" (MSc thesis, University of Iceland, 2009).

4 A.E.S. Jennings, S. Hagen, J. Harðardóttir, R. Stein, A.E.J. Ogilvie, and I. Jónsdóttir, "Oceanographic Change and Terrestrial Human Impacts in a Post AD 1400 Sediment

Record from the Southwest Iceland Shelf," *Climatic Change* 48 (2001): 83-100; Geirsdottir et al., "A 2000 Year Record of Climate Variations."

5 Árni Daníel Júlíusson, "Valkostir sögunnar: um landbúnað fyrir 1700 og Þjóðfélagsþróun á 14.-16. öld," *Saga* 36 (1998): 77-111.

6 I.A. Simpson, A.J. Dugmore, A. Thomson, and O. Vesteinsson, "Crossing the Thresholds: Human Ecology and Historical Patterns of Landscape Degradation," *Catena* 42 (2001): 175-92.

7 Árni Daníel Júlíusson, Helgi Skúli Kjartansson, and Jón Ólafur Ísberg, *Íslenskur söguatlas* [Icelandic historical atlas], vol. 2, *From 18th Century to Independence* (Reykjavik: Iðunn, 1992).

8 Sveinbjörn Blöndal, *Sauðasalan til Bretlands* [The Icelandic sheep trade with Britain] (Reykjavik: University of Iceland, Háskóli Íslands, Institute of History, 1982).

9 T. Thordarson and G. Larsen, "Volcanism in Iceland in Historical Time: Volcano Types, Eruption Styles and Eruptive History," *Journal of Geodynamics* 43 (2007): 118-52.

10 For further details, see ibid.

11 O. Arnalds, E.F. Þórarinsdóttir, S. Metúsalemarsson, Á. Jónsson, E. Grétarsson, and A. Árnason, *Soil Erosion in Iceland* (Reykjavik: Soil Conservation Service and Agricultural Research Institute, 1997).

12 Sigurður Þórarinsson, "Uppblástur á Íslandi í ljósi öskulagarannsókna," *Ársrit Skógræktarfélags Íslands* (1961): 17-54; Jennings et al., "Oceanographic Change and Terrestrial Human Impacts"; Larsen et al., "A 3000-Year Varved Record."

13 O. Arnalds, A.L. Aradottir, and I. Thorsteinsson, "The Nature and Restoration of Denuded Areas in Iceland," *Arctic and Alpine Research* 19 (1987): 518-25.

14 Larsen et al., "A 3000-Year Varved Record."

15 Ibid.

16 Ibid.; Þórarinsson, "Uppblástur á Íslandi í ljósi öskulagarannsókna."

17 Jennings et al., "Oceanographic Change and Terrestrial Human Impacts."

18 Þorvaldur Thoroddsen, *Árferði á Íslandi í þúsund ár* (Copenhagen: Hið íslenska fræðafélag, 1916-17); M.E. Mann and P.D. Jones, "Global Surface Temperatures over the Past Two Millennia," *Geophysical Research Letters* 30 (2003): 1820-23; Jennings et al., "Oceanographic Change and Terrestrial Human Impacts"; Larsen et al., "A 3000-Year Varved Record."

19 Geirsdottir et al., "A 2000 Year Record of Climate Variations."

20 Jennings et al., "Oceanographic Change and Terrestrial Human Impacts"; Larsen et al., "A 3000-Year Varved Record."

21 Geirsdottir et al., "A 2000 Year Record of Climate Variations."

22 *Hagskinna*, Icelandic Historical Statistics, Hagstofa Íslands (Reykjavik: Hagstofa Islands/Statistics Iceland, 1997).

23 Margrét Hallsdóttir, "On Pre-Settlement History of Icelandic Vegetation," *Búvísindi* 9 (1995): 17-29; Anna Guðrún Þórhallsdóttir og Björn Þorsteinsson, "Þróun búfjárhálds og gróðurfars í Hálsasveit og Hvítársíðu," *Fræðaþing landbúnaðarins*, 2005: 195-202.

24 *The Book of the Settlement of Iceland*, trans. T. Ellwood (Kendal, UK: T. Wilson, 1898), 4.

25 Sverrir Jakobsson, "Hvert er heimildargildi Landnámu? Hvenær er talið að hún hafi verið notuð?" *Vísindavefurinn*, 19 August 2000, http://visindavefur.is/.
26 O. Arnalds, "Ecosystem Disturbance and Recovery in Iceland," *Arctic and Alpine Research* 19 (1987): 508-13.
27 Hallsdóttir, "On Pre-Settlement History of Icelandic Vegetation"; Lawson et al., "Environmental Impacts of the Norse Settlement"; K. Vickers, E. Erlendsson, M.J. Church, K.J. Edwards, and J. Bending, "1000 Years of Environmental Change and Human Impact at Stóra-Mörk, Southern Iceland: A Multiproxy Study of a Dynamic and Vulnerable Landscape," *Holocene* 21 (2011): 979-95.
28 Hallsdóttir, "On Pre-Settlement History of Icelandic Vegetation"; Lawson et al., "Environmental Impacts of the Norse Settlement"; Vickers et al., "1000 Years of Environmental Change."
29 On the increase in grasses and sedges and decline of woodland species, see Hallsdóttir, "On Pre-Settlement History of Icelandic Vegetation"; Vickers et al., "1000 Years of Environmental Change."
30 S.A. Zimov, "Pleistocene Park: Return of the Mammoth's Ecosystem," *Science* 308 (2005): 796-98.
31 Páll Bergþórsson, "Hitafar og gróður," *Búvísindi* 10 (1996): 141-64.
32 Ibid.
33 B. Myhre, *Norges Landbrukshistorie I. 4000 f.Kr.-1350 e.Kr. Jorda blir levevei.* Del 1. Landbruk, landskap og samfunn 4000 f.Kr. – 800 e.Kr. [Norwegian agricultural history I: 4000 f.Kr-1350 e.Kr; The land as basis for living, vol. 1, Agriculture, landscape and society] (Oslo: Det Norske Samlaget, 2002).
34 Júlíusson, "Valkostir sögunnar."
35 Bragi Sigurjónsson, *Göngur og réttir* [Sheep gathering in Iceland] (Reykjavik: Skjaldborg Akureyri, 1987).
36 Þorvaldur Thoroddsen: *Lýsing Íslands Þriðja bindi. Landbúnaður á Íslandi. Sögulegt yfirlit. 1. bindi* [The description of Iceland, vol. 3, Agriculture in Iceland: An historical review] (Copenhagen: Icelandic Literary Society, 1919), 187.
37 Bragi Sigurjónsson, *Göngur og réttir.*
38 *Tölfræðihandbók* 1984, *Statistical handbook*, Reykjavik.
39 *Hagskinna.*
40 A shieling (*sel* in Icelandic, *sæter* in Norwegian) is a summer farm. It is located away from the home farm, on its land, often in higher ground. At the shieling, the cattle and sheep are grazed during the summer and milked, with the milk then processed into butter, cheese, and whey. Traditionally, three women worked at each shieling in Iceland. No operating shielings are extant in Iceland, unlike in Norway. Þorvaldur Thoroddsen, *Lýsing Íslands Þriðja bindi*, 207-13.
41 Ibid., 322-28.
42 Jón Steffensen, *Menning og meinsemdir: ritgerðasafn um mótunarsögu íslenzkrar þjóðar og baráttu hennar við hungur og sóttir* [Culture and lesions: Essays on the shaping of the Icelandic nation and the battle against hunger/starvation and diseases] (Reykjavik: Historical Society, 1975).
43 Jason W. Moore, "'Amsterdam Is Standing on Norway' Part I: The Alchemy of Capital, Empire and Nature in the Diaspora of Silver, 1545-1648," *Journal of Agrarian*

Change 10 (2010): 33-68; Jason W. Moore, "'Amsterdam Is Standing on Norway' Part II: The Global North Atlantic in the Ecological Revolution of the Long Seventeenth Century," *Journal of Agrarian Change* 10 (2010): 188-277.

44 Jason W. Moore, "Sugar and the Expansion of the Early Modern World-Economy: Commodity Frontiers, Ecological Transformation, and Industrialization," *Review* (Fernand Braudel Center) 23 (2000): 409-33.

45 Immanuel Wallerstein, *The Modern World System: Capitalist Agriculture and the Origins of the European World-Economy in the Sixteenth Century* (New York: Academic Press, 1974).

46 Moore, "Sugar," 410.

47 Ibid., 412-13, 410.

48 Ibid., 412-13.

49 Jared Diamond, *Collapse: How Societies Choose to Fail or Succeed* (New York: Penguin, 2005); Joakim Radkau, *Nature and Power: A Global History of the Environment* (Cambridge: Cambridge University Press, 2008).

50 Jón Sigurðsson, *Lítil varningsbók* (Reykjavik: Thiele, 1861).

51 *Hagskinna*; Júlíusson, Kjartansson, and Ólafur Ísberg, *Íslenskur söguatlas.*

52 Árni Daníel Júlíusson and Jón Ólafur Ísberg, *Íslandssagan í máli og myndum. Ný og uppfærð útgáfa af Íslenskum söguatlas* [The illustrated history of Iceland (new and updated edition of the Icelandic historical atlas)] (Reykjavik: Edda Rv, 2005); *Krambodsbøger* 1759-63. Rigsarkivet, København. Fol. 140 B [Trade reports 1759-63, Danish National Archives, fol. 140 B. Trade reports from Djúpivogur, East Iceland, 1843-45]; *Verslunarbækur Djúpavogs 1843-45,* Þjóðskjalasafn Íslands, Icelandic National Archives, Reykjavik.

53 Bragi Sigurjónsson, *Göngur og réttir.*

54 *Hagskinna.*

55 Ibid.

56 Sigurðsson, *Lítil varningsbók* [Some notes on Icelandic agriculture], 38.

57 Blöndal, *Sauðasalan til Bretlands.*

58 Lúðvík Kristjánsson, *Úr heimsborg í grjótaþorp: Ævisaga Þorláks Ó. Johnson* [From city to village: The biography of Þorlákur Ó. Johnson], vol. 1. (Reykjavik: Skuggsjá, 1962).

59 Blöndal, *Sauðasalan til Bretlands.*

60 Júlíusson, Kjartansson, and Ólafur Ísberg, *Íslenskur söguatlas.*

61 Helga Ögmundardóttir, "The Shepherds of Þjórsárver: Traditional Use and Hydro Power Development in the Commons of the Icelandic Highland," *Studies in Cultural Anthropology* 49 (PhD diss., Uppsala University, 2011).

62 Steinþór Gestsson á Hæli, "Fjárleitir og fjallkóngar í Gnúpverjahreppi," *Árnesingur* 6 (2004): 112-72.

63 Guðmundur Árnason, "Uppblástur og eyðing býla í Landsveit" [The erosion and destruction of farms in Landsveit, south Iceland], in *Sandgræðslan, minnzt 50 ára starfs* [The Soil Conservation Service – 50 years memorial], ed. Arnór Sigurjónsson (Reykjavik: Búnaðarfélagið og Sandgræðslan, 1958), 50-87.

More Things on Heaven and Earth
Modernism and Reindeer in Chukotka and Alaska

BATHSHEBA DEMUTH

On a modern map, the shoulders of Eurasia and North America nearly touch at the Bering Strait, an eighty-three-kilometre barrier between Old World and New (Map 8.1). During the rolling period of the Ice Ages known as the Pleistocene, the Pacific Ocean pulled back, leaving the Chukchi Peninsula connected to Alaska's Seward Peninsula by a wide, grassy plain. Two million years ago, *Rangifer tarandus*, the animal known as the reindeer in Eurasia and as caribou in North America, emerged along this continental juncture.[1] As products of the Ice Age, reindeer are adapted to not just survive but also thrive in million-strong herds in the dark and cold. Like all living things in the North, they must solve the problem of energy. With the sun gone for months of the year, the photosynthetic transfer of heat into palatable calories is minimal; northern plants are small and tough. But reindeer can metabolize even rocklike lichens. For humans, who can eat but not survive on moss and tree bark, and who must borrow the hides of tougher animals to withstand cold, reindeer make the tundra habitable.[2] Reindeer have been eaten by people for as long as our species has lived in northern Asia and North America, some fifteen to twenty thousand years.[3]

In our pursuit of survival, human beings are capable of adaptation; we learn from experience and communicate that knowledge between individuals and across generations without modifying our genes. As people have used reindeer in order to survive in the Arctic, adaptation has produced a series of technologies – from the most basic recognition of *Rangifer* as a

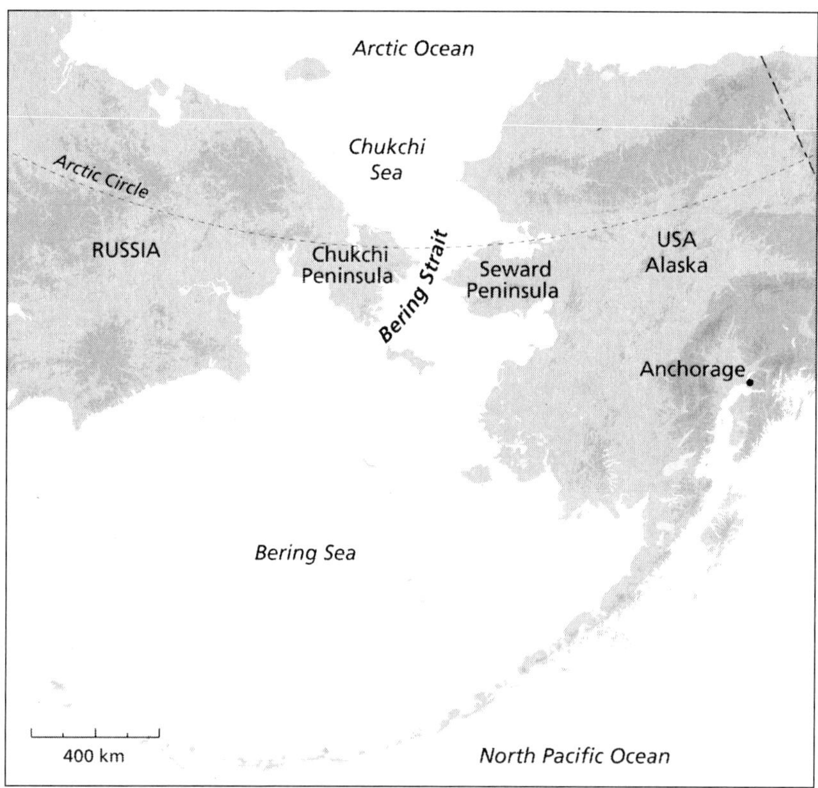

Map 8.1 Bering Strait region, including Chukotka and Alaska

source of food and hides to increasingly sophisticated methods of accessing these resources. About eight thousand years ago, people living in Chukotka hunted reindeer on skis, with canine assistance.[4] Three thousand years later, another group brought new tools to the New World: bows and arrows tipped in tiny stone points that made hunting considerably more efficient.[5]

Among the technologies that shaped interactions between humans and reindeer, one was not about killing the animals but about keeping them alive. Perhaps a thousand years ago, hunters began to breed reindeer that could be handled and herded, making the reindeer itself something deliberately altered by humans, a technology in its own right.[6] The early history of domestication is dim, but over several centuries the practice reached the Chukchi in the Far East. Unlike other innovations that crossed into North America from Chukotka, reindeer husbandry never left Eurasia. In the seventeenth century, Russians found the Chukchi herding reindeer in the interior and killing sea mammals on the coast; when the Russians crossed to Alaska in the eighteenth century, they met Inupiat hunting wild caribou and whales.

The Europeans who came to the North Pacific, first eastward from the Russian Empire and then westward from British Canada, were not aware that herding among the Chukchi was only a few centuries old, or that the ancestral Inupiat came to the New World not long before the birth of Christ.[7] The Europeans instead focused on a culture that seemed to them profoundly strange, prone to such unexpected behaviour as sharing wives, cleaning bowls with saliva, and leaving the dead unburied.[8] Whether these habits were interpreted as romantic or barbaric depended on the observer, but the strangeness of the Inupiat and Chukchi left them out of joint with European history – childlike, sometimes warlike, but changeless and primitive. For the states that came to control the North Pacific, both the space and its inhabitants posed a problem: What to do with an unchanging, cold land populated mostly by people who possessed "no conception of time or ages"?[9]

In the last decade of the nineteenth century, the United States began to grapple with these questions, joined thirty years later by its new Soviet neighbour. The timing was not accidental. The United States had grown to define itself, as Frederick Jackson Turner famously expressed, through the process of taking wild land and turning it into farms and cities.[10] With the closing of the continental frontier in 1890, Alaska began to look like the last space where this critical part of the national character could be enacted. The Soviet Union, meanwhile, inherited from the Russian Empire its substantial territories but not its limited aspirations toward the North: in the new revolutionary state, where everyone was meant to be materially equal and ideologically dedicated, Marxism could not stop at the tundra's edge.

Both countries were motivated by a sense that the North needed to become a part of their unique future; whether a socialist one or a democratic capitalist one, it was a future understood by its promoters as fundamentally better than the present. In both cases, the Inupiat and Chukchi were seen as living in a condition very distant from that future, frozen out of civilization and history itself. Such judgments were levelled, to a large degree, because of indigenous peoples' relationship with northern space.

For Americans looking north, history was a thing possessed by individuals who transformed land from a wild state to one of tamed productivity. The act of transformation was the basis of ownership; land was unowned as long as it was unimproved, and improvement meant, in essence, producing a surplus. Once it became private property, land was a source of wealth and the guarantor of political freedom; property meant both economic and political independence. By this judgment, the Inupiat, who had limited private property, seemed to exist outside the potential for progress and change.

The Soviet Union, meanwhile, found the Chukchi to be not so much without time as living in the wrong one. If Marxist history was a progression from feudalism to capitalism to socialism and onward to utopia, the nomads of the far northeast were stuck in a past where human life was unrelentingly restricted by nature. Despite herding reindeer, the Chukchi lacked enough control over the material world to achieve more than bare survival, a condition that for Marxists also restricted development toward communal freedom, a state freed from the linked evils of material worry and political oppression. For the Soviets, the Chukchi existed in a present that was really part of the deep past.

The solution to being either timeless or backward had to do with production. As long as the peoples of the North did not produce a surplus, they would remain under the boot heel of necessity and their lives could not progress toward a better future. The introduction of history required altering the relationship between land and people; making a place hopeless for traditional agriculture, difficult for industry, and often challenging to basic survival contributes not just to the subsistence of a few Aboriginal people but also produces a surplus for the betterment of the general citizenry. Through such production, the peoples of the North would escape what seemed to be, in the eyes of American and Soviet observers, lives of deprivation and exploitation.

Efforts to change Inupiat lives began a quarter century before the Soviet Union initiated its own and considerably more substantial efforts among the Chukchi, but for American and Soviet reformers alike, *Rangifer tarandus* was the key to bringing the northern wastes and their peoples into history.[11] For both countries, the means and the ends of reindeer husbandry shared a broadly high-modernist cadence: through scientific innovations and educational intervention, production would expand, human control over nature would increase, and as a result people would live fundamentally better lives.[12] There were considerable and consequential differences in the particulars, but on both Bering coasts, progress was a considered a fact and technology was the means to this better end. The ability to command nature was a given. That the reindeer might not cooperate with human attempts to control them was incongruent with high modern ideals, which in the twentieth century reshaped the relationship between people and reindeer in the North Pacific world.

The Booming West: Alaska, 1890-1920

From a ship's deck in the Pacific, the right shoulder of the Bering Strait emerges from the fog as "a desolate, moss-grown plain" where "no tree, no shrub delights the eye."[13] Uninviting as the Seward Peninsula appeared to Europeans in the late nineteenth century, it supported diverse fauna – whales, walrus, seals, and fish along the coast; caribou, Dall sheep, and small game on the tundra, itself rich in lichens, mosses, grasses, and bushes. The land produced enough food for the indigenous population to live in semi-permanent villages; the general Inupiat adaptation to their environment was intermittent mobility – ranging across a large territory when game was scarce, or settling when calories were plentiful.[14]

In 1890, Sheldon Jackson, Alaska's first commissioner of education, sailed up the west coast to visit the Inupiat villages under his jurisdiction, with a brief trip across the strait to Chukchi country. Jackson had come to Alaska initially as a Presbyterian missionary; after founding several mission schools, he lobbied for federal funds to develop the region's educational system and was subsequently appointed oversight of its development. As commissioner, Jackson saw a crisis along the coast: Inupiat were starving – their traditional livelihood threatened by European whalers and breach-loading rifles, leaving their culture in inevitable decline.[15] Yet, Jackson believed that although the Inupiat as Inupiat were doomed, the Inupiat as potential citizens were not. The key was to find something that would keep native body and soul together without resorting to the degradations of government charity. Jackson found his answer among the Chukchi, who seemed well fed and comparatively civilized. The key to Chukchi success, in Jackson's estimation, was the domesticated reindeer. If imported to Alaska, Jackson argued, these animals could "civilize, build up [Inupiat] manhood, and lift them into self-support" while also transforming Alaska into a productive space, making "millions of acres of moss-covered tundra conducive to the wealth of the country."[16] Moreover, the reindeer was an animal that created legal and social relations "like any other owner and property."[17] As property, they made the native population independent from both nature and assistance, leaving them free to become full American citizens. These new citizens would take the "useless and barren wastes" of Alaska and make them "conducive to the wealth and prosperity of the United States."[18] Reindeer would turn hunters into Jeffersonian yeoman herders and thus into Americans, and Americans would turn the tundra from waste to modern productive space.

Figure 8.1 Cape Prince of Whales reindeer herders. Perry D. Palmer Photograph Album, c. 1903-13, UAF-2004-120-28. Courtesy of University of Alaska Archives, Fairbanks.

Making Alaska wealthy and productive turned on two points: there needed to be reindeer, and the reindeer needed to be owned and managed efficiently. These objectives, in the first decade of Jackson's project, were interrelated; reindeer had to be shipped from Russia and protected so the herd would increase, and protection required knowledge. In 1891, having raised enough cash to fund a trip to Russia, Jackson brought 171 animals across the strait, along with four Chukchi to teach the Inupiat reindeer husbandry.[19]

Jackson was quite convinced that the Inupiat were starving. It seems unlikely that the Inupiat, at least in the region where Jackson first introduced reindeer, would have been in agreement. Certainly, the sea mammal population was under considerable pressure from British and American ships, and the caribou were experiencing a periodic natural decline, but there was still meat and fish to be eaten.[20] Thus, when Jackson established the Teller Reindeer Station at Port Clarence in 1892, he was addressing a crisis unseen by its supposed victims. What the Inupiat did see were the Chukchi, a group with whom native Alaskans sustained both a long-term trade relationship and considerable enmity. The Inupiat also saw the possible loss of their lucrative monopoly over the spotted reindeer hides traded out of Chukotka. It was not an ideal social climate in which to persuade young Inupiat men to move to the Teller station and take orders from old enemies and new missionary staff.

The Teller station did have the lure of reindeer, however: the Inupiat were promised animals upon completing a term as apprentices. Initial interest came from prominent Inupiat families, who attained their position by amassing and redistributing wealth. To earn a herd, and augment the networks of patronage that kept Inupiat social life together, apprentices had to relocate to the Teller station and adopt European habits from missionaries and herding skills from the Chukchi. Chukchi methods, however, were often perturbing to the Teller's European staff. Herding was supposed to civilize the Inupiat, but what if they adopted the habit of drinking reindeer milk directly from the doe's udder, or started to eat warble fly larvae like the Chukchi? By the end of the first year, Jackson was looking to recruit replacement teachers: Scandinavian Sami, who were often literate, Christian, and ideal northern farmers, according to Jackson. By the spring of 1893, Jackson's tireless lobbying secured the reindeer project federal support, and Jackson was able to bring sixteen Sami men from Norway to Alaska. The American press, covering their journey westward from New York, heralded the new instructors as "civilized Natives" who had created a "vast commercial industry" in Europe; their influence would hopefully guide the Alaskan land and its natives toward the same future.[21]

The vast commercial industry envisioned by Jackson required vast herds, and vast herds required that humans adapt to reindeer needs. Domestic reindeer might have been a technology themselves, but they were not static objects: if left alone, the imported Siberian animals wandered into avalanches, broke bones, or were eaten by wolves. Fawns froze or suffered from insects. Untended stock roamed beyond human reach or joined wild caribou herds. Husbandry thus required constant vigilance, which in turn necessitated an existence more transient than most Inupiat preferred. As a result, the social adjustments required to tend reindeer often dissuaded Inupiat apprentices. The difficulty in keeping human minders did not, however, prevent the reindeer from reproducing; by 1897, domesticated *Rangifer* numbered over two thousand, scattered at five reindeer stations.

Jackson saw the herds' growth as work of national importance – a view that gained more traction in Washington after the gold strike at Nome in 1898.[22] The tide of miners who came north in the following rush brought with them the sense that Alaska was on the road to being more than just a cold backwater. Reindeer were suddenly in demand for their meat and as draft animals. Who should best own and manage the reindeer, however, was in question during the gold boom and in the decades following. Alaska's Bureau of Education, which inherited the reindeer project upon Jackson's retirement, pushed for "the successful commercializing of the industry" among Inupiat herders and facilitated the creation of joint stock corporations that pooled the reindeer of individual Inupiat owners into large, economically viable herds.[23] Yet, there were more reindeer than there were Inupiat able and willing to tend them, and more reindeer than the government could control – over seventy thousand head by 1915.[24]

In 1914, Carl Lomen, an entrepreneur based in Nome, bought twelve hundred reindeer from a Sami herder in Alaska, with the ambition of creating a large-scale industry – something that, in his eyes, the indigenous herders were failing to achieve. Lomen's motives and those of the Bureau of Education were not vastly different: both believed that Alaska was supposed to feed people, and a lot of them. For non-natives looking at the North, this would make it modern by combating the whims of nature that produced starvation and by tying people to the rest of the country through the market. The industry seemed poised to make of the tundra something great. In 1921, having eaten reindeer meat in Nome, the Arctic adventurer John Burnham observed that the world's population would soon consume tons of reindeer, and such a future was "a commonplace statement of the inevitable."[25]

Revolutionary East: Chukotka, 1917-40

When Burnham wrote of the inevitability of the tundra feeding the world, he was sailing to Chukotka, a country "of virgin loveliness."[26] The peninsula was, of course, hardly virgin to its long-time inhabitants, the Chukchi and coastal Yupik. It also stood within the reach of revolutions and all the history they entail. However, in 1921, the revolution on the ground in Chukotka was not yet Bolshevik but rather came from the growth of reindeer pastoralism. The eighteenth century was a cool time in the Arctic, which was good for reindeer. The Chukchi, who had used domesticated reindeer as draft animals for a century but avoided killing them for food, suddenly had a surplus of tame animals. For the next fifty to eighty years, the domestic herds grew, while the wild reindeer hunt continued. Then, in the early nineteenth century, the climate warmed and the reindeer population, both wild and domestic, decreased. Driven by a need for meat and hides, herders began to kill their stock. Domestic reindeer transformed from a technology of transport to one that provided food as well, a shift that allowed the human population to quadruple in four or five generations.[27]

The advent of specialized, nomadic reindeer herders produced a self-perpetuating surplus of calories and hides. With surplus came social stratification; not all herds were the same size, and not all Chukchi owned reindeer – making Chukotka in the early twentieth century home to considerable inequality, with 5 to 10 percent of the herders controlling between half and two-thirds of the reindeer.[28] It was, in Marxist terms, a society of "rich exploiters" who left the poorest to bear "the full burden of colonial and class oppression."[29]

Soviet reformers saw class exploitation among the Chukchi, much like American missionaries saw starvation among the Inupiat, as a social problem; as in America, the Soviet Union found a solution in the human relationship with reindeer. In Chukotka, the socialist form of modernity required taking privately owned herds and merging them into a collective enterprise that would uplift the exploited herder. Moreover, putting the "backward tribes of the north" on the road to socialism was, in the argument of Chukchi ethnographer Waldemar Bogoras, not merely an act of class solidarity but of national importance, since almost one-third of Soviet territory was underexploited taiga and tundra.[30] To transform the northern wastes and their people would require prompting: Bogoras argued that "we must send to the North not scholars but missionaries, missionaries of the new culture and new Soviet statehood."[31]

It would take some time for revolution to break over Chukotka. The czarist government maintained a tenuous hold among the Chukchi, leaving little organizational capacity for the Soviets to requisition.[32] When the representatives of the revolution finally arrived in 1923, they were "first of all struck by the backwardness of the [Chukchi] population."[33] The first Soviets in the Far East were as disconcerted by the Chukchi as the Americans employing them in Alaska had been: both were particularly appalled by the eating of lice and habit of using the same bowls for human waste and human food.[34] It was possible to be too primitive a Communist. Education, indeed transformation, was required.

This was a delicate process, however. The few Soviets who came to Chukotka in the 1920s and 1930s had missionary zeal but did not speak Chukchi or possess the skills to contend with the "severe winter, which lasts almost all year long."[35] Many came prepared to find the proto-Communist societies described by Marx but instead were greeted by a social order that tolerated considerable inequity. Equally delicate was the question of reindeer. The fight against backwardness required making the Arctic landscape more productive and more equal. By the late 1920s, this meant collectivization, the creation of state farms (*sovkhozy*) and collective farms (*kolkhozy*). In both cases, small-scale herders were to pool their deer, and the rich "exploiting" class, or kulaks, would relinquish their stock for the common good. The purpose was dramatic social and economic change: collective farms, in the words of Anatolii Skachko, the last head of the People's Commissariat of National Ties and a vocal advocate of northern peoples and their development, "must not just become an important source of raw materials" but were the chief way to stimulate "the collective action of poor and middling herders, assisting in their liberation from material dependency on the kulaks."[36]

The instigators of this transformation, however, knew very little about reindeer. Ivan Druri, who arrived in Chukotka in 1929, recalled how the first step toward collectivization was getting "the advice and help of the local experienced herders."[37] Druri, a committed first-generation Communist intent on bringing Lenin's word to the North, spent the next three years building the Snezhnoe *sovkhoz,* a process that began with "regular first-hand observations in the nomad camps," where Druri and his colleagues studied reindeer production.[38] The Communist emissaries noticed that Chukchi with the largest herds maximized the ratio of females to males, which kept the annual meat production high. Small herds had little leftover meat or hides after household use; only large numbers of animals offered a commercially significant surplus.[39] Since surplus was the goal, Druri and other Soviets in the North concluded that state and collective farms needed large herds.

While Druri was learning from the Chukchi, the Chukchi were learning what, exactly, collectivization meant: relinquishing their private property to the control of the state. Druri initially tried to buy deer from wealthy herders, but as the ideological climate of the 1930s increased the pressure to collectivize across the country, purchase was replaced with outright seizure. In response, wealthier Chukchi chose to burn the tundra and slaughtered their reindeer herds en masse rather than hand their animals over to the state.[40] Other herders moved deeper into the tundra. "The reindeer has many enemies," one *kolkhoz* director wrote, and "much effort and persistence is required of herders to protect and increase the stock of reindeer."[41] In the chaos of collectivization, there was not sufficient Chukchi persistence to go around or non-native expertise to compensate. Revolution became another enemy of the reindeer, at least in tamed form: the Chukotkan herds lost over a hundred thousand animals before 1940.[42]

Soviet planners in the North were aware that having the Chukchi kill their own stock was hardly furthering the goals of socialism. While full collectivization – the commitment to which was a sort of loyalty test for rural officials across the Soviet Union in the 1930s – was still the goal, the pressure to collectivize every living thing and practical object on the tundra was reduced. Families were allowed to privately own up to six hundred animals. These changes eased the resistance to collective herding, and by 1940 over 80 percent of Chukotkan reindeer were running in collective herds managed by dozens of *kolkhozy* and a few state farms.[43]

The presence of numerous small collectives did not signal the end to the perceived backwardness of the Chukchi, especially when it came to education. Schools were constructed around the peninsula starting in the 1920s, but nomadic Chukchi parents had to leave their children with non-native teachers for months at a time, leading to low enrolment and many runaways.[44] School required settlement, but reindeer had to move to find food. By the 1930s, an emerging group of reindeer specialists proposed a solution to the simultaneous need for settlement and transience. With proper, technologically advanced, large-scale collectivization, the peripatetic work of herding could be shifted from many families of varied wealth and uniform misery to specialized brigades that "would liberate women and children from the burdens of nomadic life."[45]

Collectivization was also supposed to solve the reindeer problem. In the era of Five-Year Plans, surpluses from every industry were supposed to become more astonishing regardless of demand. Yet, the herd numbers were not cooperating – in the 1930s and 1940s, the head count was shrinking.[46] For the engineers of northern advancement, the answer to low numbers was not related to the biological condition of their herds so much as to the herd's form: more animals would be the result of consolidation, since large herds produced the most meat and could be minded by specialized brigades, which in turn facilitated better oversight of the animals and pastures. The Soviets wanted to create a new environment: the tundra as a highly productive space, demonstrating the power of socialist progress.

The Seward and Chukchi Peninsulas, 1930-60

While the Soviet reindeer spent the 1920s and 1930s being reordered, reimagined, and reduced, the herds in Alaska were flourishing, reaching 400,000 head by 1929.[47] As the total number of reindeer grew, so did the size of individual herds. Following the construction of a federal reindeer experiment station at Unalakeet in 1920, the US Bureau of Biological Survey argued that small, closely watched herds run by the Inupiat collectives were inefficient; it was better to keep many animals in open groups. To increase production, the Biological Survey recommended that reindeer be treated like cattle on the western prairies, where centralized ranches allowed for settled life, while animals grazed "on a system of allotted ranges" that, with "improvement of herd management, enactment and enforcement of a brand registry law, and the control of diseases and parasites" would increase reindeer productivity.[48]

Finding a rational method of allotting range land which in the 1920s was federal, technically owned by everyone and thus, in practice, effectively owned by no one – was critical for the Biological Survey. Without ordered control over who grazed their animals where, some parts of the tundra were overused and others left fallow. L.J. Palmer, a senior biologist at the Biological Survey, based his analysis of herding efficacy on the emerging ecological concept of carrying capacity, or "the number of animals that can find sufficient palatable forage ... year after year, without injuring the plants."[49] Palmer calculated that each reindeer required thirty acres of good range, if managed so as to protect lichen growth and combat overgrazing. With proper – meaning efficient – administration of the tundra, Palmer believed Alaska's reindeer should run three to four million head.[50]

While reindeer could be "a large factor in the development of the [Alaska] Territory," who would own this development was still up for debate.[51] Leading into the Great Depression, both Inupiat cooperatives and the Lomen Reindeer and Trading Corporation had a claim on the industry. Both needed markets for what was now an undeniable surplus. To keep the peace, Carl Lomen sold reindeer products only outside Alaska, leaving local concessions to native herders. Lomen was an aggressive and creative marketer, courting food critics, pet-food manufacturers, and the armed services as potential consumers.[52] This met with some success: by 1929, almost 6.5 million pounds of reindeer products had been sold to cities in the continental United States.[53] However, reindeer concessions could not avoid the ravages of the Depression or the campaigning of the beef lobby. Lomen's enterprise lurched toward bankruptcy in the 1930s.

For the Inupiat, the demand inside Alaska was low after the gold rush, leaving indigenous herders with little market and much product. When reindeer oversight passed from the Bureau of Education to the Alaska governor in 1929, the federal government audited the industry, bringing Alaskan reindeer to the attention of Roosevelt's Indian affairs commissioner, John Collier. Collier's policies emphasized cultural pluralism and the need for indigenous peoples to engage in traditional economic practices, and were threaded with a romantic understanding of native life as an antidote to the industrial economy. Collier concluded that reindeer herding was a traditional part of Inupiat existence, which should be protected and promoted exclusively for Inupiat economic and social well-being. Non-natives, therefore, had to be removed from the reindeer business. After considerable political wrangling, the 1937 Reindeer Act bought out the failing Lomen company and made non-native ownership of reindeer illegal in order to "establish and maintain for the said natives of Alaska a self-sustaining economy."[54]

The US government had first created a tradition and then had to protect it. The new tradition, however, was subject to very modern laws: the Bureau of Indian Affairs wanted to disband large collective herds for "individual enterprisers" and their employees.[55] With the government overseeing the effective use of rangeland, herd management, and new markets for reindeer products, the herders would be motivated by "fear of losing money," resulting in an efficient, productive system in which "the number and size of herds will be limited by ... supply and demand." Thus, the laws of economics and biology would produce social self-sufficiency and "give [native] owners freedom to do what they ought with regard to the rights of others."[56] The invented tradition was that of Jackson's idealized Sami: free citizens filling grocery stores while "advancing in civilization."[57]

All reindeer were now Inupiat-owned, but not all Inupiat were owners. After three decades of apprenticeships, non-native competitors, and changing government directives, the whims of policy and the marketplace made herding look no more stable than the whims of nature. Herding was, for many Inupiat, just one possible means among many for obtaining food and cash, a part of the diversified range of activities that had always allowed for survival in the North. But Jackson's experiment had created more than part-time herders; it created a new environment: one home to a great many reindeer. Having outpaced the demand for their meat and hides, the animals spent the Depression years in a state of uncertain ownership and of little market worth, often escaping both human oversight and the dinner plate.[58] Left alone, the reindeer followed their own imperatives; they migrated, grazed, and above all, bred. By the early 1930s, there were 640,000 domestic *Rangifer* in Alaska. In their vast numbers, "Alaska's food," the territory's congressional delegate noted, was "eating up Alaska."[59]

In North America, the reindeer may have been eating the future state of Alaska, but in the Soviet Union, the state was eating reindeer to save its future. Or at least this was the idea in Chukotka, which sent 4.8 million rubles' worth of reindeer meat out of the district during the Second World War, along with tons of leather.[60] The drive to defeat Fascism was accompanied by the drive to create better collectives, which by the 1940s and the 1950s were increasingly seen as large enterprises with diminished private control of reindeer. Increased herd size "insured uninterrupted growth of reindeer breeding" and the use of improved technical equipment for herding.[61] These advances, put forward by a new generation of reindeer experts, were meant to create more reindeer. To protect against disease, animals were subject to a host of new technologies: provided with sun shades in the summer, vaccinated to "liquidate" anthrax, and doused with chemical insecticides to reduce warble flies and other parasites.[62] Because wolf predation forced herders to cluster their deer in a manner damaging to the animals and the rangeland, experts sought the extermination of wolves.

Most importantly, reindeer experts wanted to foster the "rational planning of pasturing" based on aerial surveys and hand inspections of lichens, grasses, and other plants, which were used to determine pasture allotments for collectives.[63] With each farm assigned a territory, and each territory partitioned according to which "seasonal utilization" offered the best reindeer nutrition, and each parcel rotated to avoid overgrazing, reindeer scientists planned a standard tundra, edited of inefficiencies and maximally productive.[64] This was important, since the goals of socialism made quantity "the basis of correct organization."[65] Scientists had to know, therefore, the maximum number of reindeer that each collective's allotted territory could raise, so maximum efficiency could be assured. For many Soviet scientists there existed a "reindeer carrying capacity," a fixed maximum that could be determined, as in the United States, from surveys of plant types and reindeer grazing habits.[66]

Overseeing these new pastures was a new herder, no longer part of a nomad family group, but a member of a brigade that included "a foreman, 4-5 herdsmen ... a learner, and a woman who takes care of the 'chum' (herdsmen's tent)."[67] These brigades worked on the tundra for several months at a time, taking turns monitoring deer, pastures, diseases, and predators. Reindeer work had become like factory work, run in shifts and with production quotas set nationally, "based on the projected plan of economic development" and requiring specialized classroom education along with practical field instruction.[68] Doing so not only produced better "organization of land exploitation" and "technical equipment for reindeer breeding" but also left time for the "liquidation of illiteracy."[69] A rationally organized Arctic thus also created Soviet citizens, "first-rank workers of the tundra, people of a new type, who unflinchingly and every year achieve high indices in the field of reindeer breeding."[70] And the people of a new type were presiding over growing herds. There were over 500,000 reindeer in Chukotka by 1960, slightly surpassing the population in the early 1920s.[71] Such growth was a sign, according to one author, that the Soviets had "regenerated life on the cold land and conquered the dead wastes."[72]

Conclusion

In the last years of the nineteenth century and the first half of the twentieth, the relationship between human beings and *Rangifer tarandus* was transformed on both sides of the Bering Strait. In both countries, a handful of bureaucrats, educators, missionaries, and cadres dreamed high-modernist dreams about what the North could become. In the abstract, the dream of the productive tundra had a common plot on both sides of the Bering Strait: with the modernist tools of measurement and reform, each country would transform the cycles of Arctic nature into a predictable and productive space. Reindeer were the enabling object; through them, northern people would be freed from material want and social deprivation while joining their countrymen in progress toward a better future.

The ambitions of Jackson, Druri, and the other champions of northern development did create something new under the midnight sun: the reindeer as a technology created other new technologies to support domesticated herds, and these herds altered the northern space and transformed peoples' relationship to the non-human world. Semi-sedentary Inupiat became peripatetic owners of private property, while nomadic Chukchi became settled shift-working herders of collectivized reindeer – both groups becoming citizens of states with dreams of a tundra that would feed the world. The result was a new environment, a new fusion of humans, nature, and technology in the North Pacific.

Yet, change did not necessarily yield the expected results. Some Inupiat did become herders and engaged with the market by selling reindeer commodities, but reindeer meat in the United States only briefly reached markets beyond Alaska, and reindeer did not turn the majority of Inupiat into independent yeomen or make them fully equal citizens. The Soviet Union, having roiled its agricultural production in the 1930s, had more need for meat than the United States, and more dramatically changed social life among the Chukchi in producing it. The desire to collapse the differences between urban factory work and rural tundra herding as much as possible radically, and sometimes violently, altered the form of reindeer husbandry. However, the dissolution of rich herders into collectives did not manufacture equality. By the postwar period, the Chukchi were increasingly pushed into second-class jobs, and although education did produce some local party leaders and writers, the Chukchi were never masters of their own brand of Communism.[73]

Most of all, while creating new environments neither the United States nor the USSR managed a fundamental alteration to natural history, or more precisely, to change *Rangifer tarandus* and the host of non-human forces that let them breed, thrive, and multiply or go barren, starve, and perish. Both states arrived in the Arctic convinced of the region's backwardness, its timeless position outside the flow of progress. The task of government was, essentially, to remove the state of nature and replace it with pure social will. Both the Americans and Soviets brought new technologies to the Arctic meant to do just this, to speed things up, to make men of a new type on land with new potential – starting, most obviously, with the introduction of domestic animals in Alaska and extending to the use of vaccines, pesticides, new herding methods, and education for herders. What would let the tundra escape backwardness was productivity, enabled by husbanding technologies meant to not just make Beringia more productive but make it maximally so. The ideal that informed US policy in the 1920s and 1930s and Soviet management for the duration was to breed the greatest number of reindeer by herds to the threshold of the land's carrying capacity.

In theory, this seemed like a straightforward proposition. Carrying capacity was, and is, understood as the number of animals a given parcel of land can support in perpetuity. With the tonnage each reindeer needed to eat known, and the composition of pastures surveyed, with predators subtracted – quite literally – from the equation, and diseases managed, humans were meant to be the only thing left to act upon the reindeer. *Rangifer* had been effectively isolated in an environment perfect for creating more *Rangifer*. This ideal tundra was also profoundly, and ironically, outside history: once carrying capacity was reached, unless the weather was unusually terrible or some new disease arose among the mosses, it existed in stasis. In order to make the North modern, it had to become completely timeless, absolutely unchanging.

But reindeer behave according to a more complicated calculus. In both countries, the herd numbers wavered and then expanded, reaching a high of over 600,000 in northwestern Alaska in the 1930s and nearly 600,000 in Chukotka by 1970.[74] It seemed that socialism had revived the tundra and that capitalism could produce meat from the barrens. Then the herds crashed. In Alaska, domestic reindeer were in crisis by 1940, and by 1950 there were a mere 25,000 left.[75] Across the Pacific, Soviet reindeer herds never reached the dramatic number forecast by scientists, and the herds were in decline by 1980 and continued to wane, reaching a low of 150,000 head in the early 1990s.[76] The reasons for the decline are varied and complex. In Alaska, the wolf population expanded along with the reindeer, and in periods when the herds were left untended or grazed on large open ranges, predation rates were high. It is possible that reindeer starved; many animals were pastured close to the ocean, where the Inupiat lived and markets were accessible, so the range may have been overused.[77] The domesticated reindeer also arrived at a moment when the local caribou population was at a low point, and as wild *Rangifer* returned, many reindeer went feral.[78] The collapse of the Soviet Union, which left herders without salaries or equipment, likewise left the deer exposed to wolf predation, feral wandering, and poaching. And finally, reindeer herds simply decline periodically, do so rapidly, and then regain numbers slowly.[79]

Thus, what undermined reindeer husbandry was neither simply wolves nor simply men, just as it was not only men that allowed reindeer domestication to flourish thousands of years ago. Of course, human activity played a role in increasing or introducing reindeer stock, and certainly the twentieth century saw particular technologies of reindeer use employed by modern states, particularly the standardized use of rangeland in order to reach maximum herd size. However, the herds did not solely decline because of human miscalculations, just as the herds would not increase simply to meet human calculations; even humans lack the raw force to be the only thing acting on the tundra. Models of carrying capacity were meant to develop and improve – to create history – but they did so in an ecological context that, frozen appearance aside, is always changing. With a living technology like reindeer, in a natural space like the North, it is not possible to make an environment static – to remove nature from the equation.[80] There are more things on heaven and earth than even the most perceptive state or nomad has the power to rush or remove; even if every wolf on the tundra was exterminated and every disease eradicated, the reindeer population still declines every sixty to one hundred years. High modernism is, if nothing else, a narrative of humans making history. However, in the North there is no escape from timelessness into history simply through new technology. History in the North, like history anywhere, is both deeply human and trans-human; the land itself has a past and is changing under hoof and foot.

NOTES

1 E. Anderson, "Who's Who in the Pleistocene: A Mammalian Bestiary," in *Quaternary Extinctions: A Prehistoric Revolution*, ed. P.S. Martin and R.G. Klein (Tucson: University of Arizona Press, 1984), 40-89.

2 Valarius Geist, "Of Reindeer and Man, Human and Neanderthal," special issue, *Rangifer* 14 (2003): 57-63.

3 Stuart Fiedel, "The Peopling of the New World: Present Evidence, New Theories, and Future Directions," *Journal of Archaeological Research* 8, 1 (2000): 39-103. See also John F. Hoffecker and Scott Elias, "Environment and Archeology in Beringia," *Evolutionary Anthropology* 12 (2003): 34-49.

4 N. Dikov, *Naskal'nye zagadki dreveny Chukotki: Petroglify Pegtymelya* (Moscow: Nauka, 1971).

5 Igor Krupnik, *Arctic Adaptations: Native Whalers and Reindeer Herders of Northern Eurasia*, trans. Marcia Levenson (Hanover, NH: Dartmouth College Press, 1993), 144-47.

6 Leonid Baskin, "Reindeer Husbandry/Hunting in Russia in the Past, Present and Future," *Polar Research* 19, 1 (2000): 23-29; Krupnik, *Arctic Adaptations*, 166-68.

7 Don E. Dumond, "A Reexamination of Eskimo-Aleut Prehistory," *American Anthropologist* 89, 1 (1987): 32-56.

8 Waldemar Bogoras, *The Chukchee* (New York: E.J. Brill, 1904), 480-85; Charles Madsen, *Arctic Trader* (New York: Dodd, Mead, 1957), 133-50.

9 Richard Bush, *Reindeer, Dogs, and Snow-Shoes: A Journal of Siberian Travel and Explorations* (New York: Harper and Brothers, 1871), 460. For the image of indigenous people, see Slezkine, *Arctic Mirrors*, and Robert Berkhofer Jr., *White Man's Indian: Images of the American Indian from Columbus to the Present* (New York: Vintage, 1978).

10 Frederick Jackson Turner, "The Significance of the Frontier in American History," *The Annual Report of the American Historical Association* (Madison: University of Wisconsin Press, 1893), 199-227.

11 Roxanne Willis places Sheldon Jackson's reindeer project at the beginning of a long line of development efforts in Alaska. See Roxanne Willis, *Alaska's Place in the West:*

From the Last Frontier to the Last Great Wilderness (Lawrence: University Press of Kansas, 2010), 23-47.

12 Here I am using James Scott's definition of high modernism as outlined in *Seeing Like a State: How Certain Schemes to Improve the Human Condition Have Failed* (New Haven, CT: Yale University Press, 1998), 88. In thinking about ways in which ideas about modernity are narrated, I am also influenced by the work of James Ferguson in *The Anti-Politics Machine: "Development," Depoliticization, and Bureaucratic Power in Lesotho* (Cambridge: Cambridge University Press, 1990) and Nickolai V. Ssorin-Chaikov's work on development in Siberia, *The Social Life of the State in Subarctic Siberia* (Stanford, CA: Stanford University Press, 2003).

13 Paul Niedieck, *Cruises in the Bering Sea: Being Records of Further Sport and Travel*, trans. R.A. Ploetz (London: Rowland Ward, 1909), 171.

14 Richard Stern et al., *Eskimos, Reindeer and Land*, Bulletin 59 (Fairbanks: Agricultural Experiment Station, School of Agriculture and Land Resources Management, University of Alaska Fairbanks, 1980), 11-13; Ernest S. Burch Jr., *Eskimo Kinsmen: Changing Family Relationships in Northwest Alaska* (New York: West Publishing, 1975), 17-20.

15 By 1890, the Inupiat had been exposed to Russian, British, and American trade for the better part of a century, introducing Old World diseases and alcohol, the combined effects of which had nearly halved the Inupiat population. For non-native observers, however, it was European hunting practices that caused decline.

16 Sheldon Jackson, *The Introduction of Reindeer into Alaska, Preliminary Report of the General Agent of Education in Alaska to the Commissioner of Education 1890* (Washington, DC: Government Printing Office, 1891), 4.

17 S. Jackson, *Annual Report on the Introduction of Domestic Reindeer into Alaska, 1892.* Senate Miscellaneous Document no. 22, 52nd Congress, 2nd Session (Washington, DC: Government Printing Office, 1893), 14.

18 S. Jackson, *Report of the Commissioner of Education for the Year 1894-95*, vol. 2 (Washington, DC: Government Printing Office, 1896), 1438.

19 S. Jackson, *Report of the Commissioner of Education for the Year 1894-95*, 1440.

20 Ernest Burch Jr., "The Caribou/Wild Reindeer as a Human Resource," *American Antiquity* 37, 3 (1972): 339-68; Dean F. Olsen, *Alaska Reindeer Herdsmen: A Study of Native Management in Transition*, SEG Report no. 18 (Fairbanks: Institute of Social, Economic and Government Research, University of Alaska Fairbanks, 1969), 21.

21 *The Record* (Chicago), 11 September 1894, quoted in Willis, *Alaska's Place in the West*, 31; S. Jackson, *(Fifth Annual) Report on the Introduction of Reindeer into Alaska*, 54th Congress, 1st Session, Senate Executive Document no. 111 (Washington, DC: Government Printing Office, 1896), 17.

22 S. Jackson, *(Fifth Annual) Report on the Introduction of Reindeer into Alaska*, 10-11.

23 US Bureau of Education, *Report on the Work of the Bureau of Education for the Natives of Alaska, 1914-1915* (Washington, DC: Government Printing Office, 1917), 8.

24 US Bureau of Education, *Report on the Education of the Natives of Alaska and the Reindeer Service, 1910-11*, Alaska School Service, whole no. 484 (Washington, DC: Government Printing Office, 1912), 23; US Bureau of Education, *Report on Work of the Bureau of Education for the Natives of Alaska, 1913-1914*, 7.

25 John B. Burnham, *The Rim of Mystery: A Hunter's Wanderings in Unknown Siberian Asia* (New York: G.P. Putnam's Sons, 1929), 281.

26 Ibid., 281.

27 This paragraph is a synthesis drawn primarily from Krupnik, *Arctic Adaptations*, chaps. 4 and 5.

28 Ibid., 93-95; L.M. Baskin, *Severnyi olen: upravlenie povedeniem i populiatsiiami olenevodstvo okhota* (Moscow: Tovari shchestvo nauchnykh izdanii, 2009), 182-88.

29 V.N. Uvachan, *The Peoples of the North and Their Road to Socialism* (Moscow: Progress, 1975), 45.

30 Waldemar Bogoras, "Ob izuchenii I okhrane okrainnykh narodov," *Zhizn' natsional'nostei* 3-4 (1923): 169.

31 Waldemar Bogoras, quoted in Slezkine, *Arctic Mirrors*, 159.

32 I.S. Vdovin, *Ocherki istorii i etnografii chukchei* (Moscow: Nauka, 1965), 251.

33 B.I Mukhachev, *Bor'ba za vlast' sovetov na Chukotke (1919-1923): Sbornik dokumentov i materialov* (Magadan, Russia: Magadanskoe Knizhnoe Izdatel'stvo, 1967), 62.

34 N. Galkin, *V zemle polunochnogo solntsa* (Moscow: Molodaia gvardiia, 1929), 80-82, 113-14.

35 Mukhachev, *Bor'ba za vlast' sovetov na Chukotke*, 124.

36 Quoted in Anatolii Skachko, "Stroitel'stvo olenevodcheskikh covkhozov," *Soveskii sever* 9 (1931): 153.

37 I.V. Druri, "Kak byl sozdan pervyi olenesovkhoz na Chukotke," *Kraevedcheskie zapiski* 16 (1989): 7.

38 Druri, "Kak byl sozdan," 7.

39 Baskin, *Severnyi olen*, 185-86; I. Druri, *Olenevodstvo* (Moscow: Izdatel'stvo selkhoz literatury, zhurnalov i plakatov, 1963), 39.

40 F. Ia. Gul'chak, *Reindeer Breeding*, trans. Canadian Wildlife Service (Ottawa: Department of the Secretary of State, Bureau for Translations, 1967), 32.

41 V. Kozlov, "Khoziaictvo idet v goru," in *30 let Chukoskogo natsional'nogo okruga* (Magadan, Russia: Magadanskoe Knizhnoe Izdatel'stvo, 1960), 66.

42 Patty Gray, "Chukotkan Reindeer Husbandry in the Twentieth Century," in *Cultivating Arctic Landscapes: Knowing and Managing Animals in the Circumpolar North*, ed. David Anderson and Mark Nutall (New York: Berghahn Books, 2004), 143.

43 Ibid.

44 T.Z. Semushkin, *Children of the Soviet Arctic* (London: Travel Book Club, 1947), 35; T.Z. Semushkin, "Oryt raboty po organizatsii shkoly-internata chukotskoi kul'tbazy DVK," *Soveskii sever* 3-4 (1931): 177-82.

45 Anatolii Skachko, "Problemy Severa," *Soveskii sever* 1 (1930): 26.

46 Gray, "Chukotkan Reindeer Husbandry," 143.

47 Olsen, *Alaska Reindeer Herdsmen*, 14.

48 Seymour Hadwen and Lawrence Palmer, *Reindeer in Alaska* (Washington, DC: Government Printing Office, 1922), 69.

49 Lawrence Palmer, *Raising Reindeer in Alaska*, US Department of Agriculture Miscellaneous Publication no. 207 (Washington, DC: US Department of Agriculture, 1934), 23.

50 Hadwen and Palmer, *Reindeer in Alaska*, 4.

51 Ibid., 70.
52 Carl J. Lomen, *Fifty Years in Alaska* (New York: David McKay Company, 1956), 99.
53 Olsen, *Alaska Reindeer Herdsmen*, 14.
54 75th Congress, 1st Session, 15 and 22 June 1937 (Washington, DC: Government Printing Office, 1937), 2.
55 J.S. Rood, "Narrative re: Alaska Reindeer Herds for Calendar Year 1942, with supplementary data" (Nome, AK: US Bureau of Indian Affairs, 1943), 151.
56 Ibid., 151.
57 75th Congress, 23.
58 Olsen, *Alaska Reindeer Herdsmen*, 14.
59 Judge James Wickersham, quoted in Willis, *Alaska's Place in the West*, 41.
60 N.N. Dikov, *Ocherki istorii chukotki s drevneiskikh vremen do nashikh dnei* (Novosibirsk, Russia: Nauka, 1974), 252-53.
61 Gul'chak, *Reindeer Breeding*, 32, 191.
62 Ibid., 192.
63 Ibid., 82.
64 P.S. Zhigunov, ed. *Reindeer Husbandry*, trans. Israel Program for Scientific Translations (Springfield, VA: US Department of Commerce, 1968), 187.
65 Gul'chak, *Reindeer Breeding*, 191.
66 Zhigunov, *Reindeer Husbandry*, 3.
67 Ibid., 82.
68 Ibid., 87.
69 Gul'chak, *Reindeer Breeding*, 259-60.
70 Ibid., 260.
71 Gray, "Chukotkan Reindeer Husbandry," 143.
72 Tikhon Semushkin (Syomushkin), *Alitet Goes to the Hills* (Moscow: Foreign Languages Publishing House, 1952), 12.
73 For a discussion of social conditions in the late Soviet and post-Soviet periods, see Patty Gray, *The Predicament of Chukotka's Indigenous Movement* (Cambridge: Cambridge University Press, 2005), 103-16 and 131-53.
74 Gray, "Chukotkan Reindeer Husbandry," 143; Olsen, *Alaska Reindeer Herdsmen*, 14.
75 Olsen, *Alaska Reindeer Herdsmen*, 14-15.
76 Gray, "Chukotkan Reindeer Husbandry," 143.
77 Stern et al., *Eskimos, Reindeer and Land*, 37.
78 David R. Klein, "Limiting Factors in Caribou Population Ecology," special issue, *Rangifer* 7 (1991): 30-35.
79 Igor Krupnik, "Reindeer Pastoralism in the Time of the Crash," *Polar Research* 19, 1 (2000): 49-56.
80 Some ecologists see the local contexts in the Arctic as so variable that they are nearly impossible to document accurately. See Atle Mysterud, "The Concept of Overgrazing and Its Role in the Management of Large Herbivores," *Wildlife Biology* 12, 2 (2006): 129-41.

A Touch of Frost
Gender, Class, Technology, and the Urban Environment in an Industrializing Nordic City

SIMO LAAKKONEN

Sniffles, suffering from the cold, and occasional frostbites were part of everyday life in every Nordic city.[1] In the nineteenth century, winter in Helsinki lasted for five to six months and summer from three to four months, with very short spring and autumn transitional periods, depending on the year. The winter months normally lasted from November or December until March or April. The average January temperature was –7.2°C and it seldom dropped below –25-30°C during ordinary winters.[2] In Helsinki and other Nordic cities, winter holds a particular place among seasons.

But upon closer examination, it becomes clear that people were not in an equal position with regard to winter: Some groups were not as exposed to cold weather as others. Some people could choose whether they went out or stayed in, whereas others did not have this choice. Social differences within the population exposed some groups of people to bad weather more frequently than others. Generally, members of the upper class and most of the small middle class worked indoors, whereas the working class worked outdoors part or all of the time. However, one specific group was exposed to winter in particular, and more frequently than any other group in the city:

> Every one of us who has, during the winter or the cold, windy, and rainy spring or on autumn days, seen the poor women of the serving class, in soaking wet clothes, their limbs blue with cold, wash and beat laundry at the so-called jetty by the shore, is inevitably astounded by this barbarism.[3]

In all social classes, laundry was the responsibility of women. Because of the heavy nature of the work, those few women who could afford to do so hired other women to do their washing for them. Hence, although most women in the city did laundry work, servants and professional washerwomen carried the heaviest load.

This chapter focuses on women in the urban environment. How else can we understand the relation between our physical environment, half of the human population, and society in general? Linking women's history and environmental history is a challenging but necessary task.[4] Environmental historians have, however, largely neglected gender issues, and gender historians have failed to explore environmental issues.[5] Fortunately, some pioneering studies have been conducted, primarily in the United States, on the role of women in the urban environment. Gender and environmental historians have shown that middle- and upper-class women were active in public health and sanitation issues in a number of American cities during the Progressive Era.[6] But what about women in Europe and, more specifically, in the Nordic countries? Did Nordic women play a similar role in urban environmental issues during this era, when their sisters in the United States were so active?

Few people today believe that women inherently provide a better model for relating to nature than men. Rather, it is necessary to explore women's history with regard to specific relations that women had or may have had with their environment. In this chapter, I explore the history of laundry work, which has been the provenance of women around the world. As historian Arwen Mohun expresses it, "Whether in manor houses or army camps, by streamsides or in tenement rooms, laundrywork has traditionally been one of the most powerfully gendered of all domestic tasks."[7] Women have been responsible for this hard work, which has been one of the most important factors contributing to human well-being worldwide. Yet, it seems that the laundress and her work were too commonplace, too rough, and too undramatic to attract much attention from the greater public in the past or from historians today.[8]

This chapter investigates women and their work during a particular season – winter. What was the relation between winter and the urban society in the past, and how did winter affect different groups of people? Historical studies have provided few answers to such questions. In historical climatology, social crisis brought about by climatic fluctuations and changes, such as the Little Ice Age, has been an important theme, but most studies have focused on rural societies and agriculture in particular.[9] An increasing number of studies of the history of ice and snow have been published, but they have paid limited attention to towns and cities.[10] Some studies in urban history have addressed such issues as the impact of winter on the city and removal of snow, but much remains to be done.[11] Winter is still one of the major white spots in the historiography of the towns and cities.

My intent is to study the historical relationship between the urban environment and the women of Helsinki by wearing an apron around town. Various technologies were used or proposed to be used in laundry work, and these had or could have had impacts on working conditions. The historical relationship between laundry work and the urban environment is best understood through identity politics, meaning both self-conception and self-expression, as well as group expression and affiliation.[12] Because I concentrate on the era of industrialization, I also regard it as necessary to include the notion of class within the concept of identity. Class is a crucial angle to consider when studying women's work and women belonging to different social categories in particular.[13] What were the relations between laundry-work technology,[14] female identity, and the environment in an industrializing Nordic city? I claim that in addition to battling urban nature and the environment, washerwomen had frosty relations with the leading city fathers and cold, if not icy, relations with the leading female figures of the better-off classes of the city.

Shelter from the Storm

During the cold part of the year, laundry was washed indoors in kitchens, saunas, or washhouses, where it was possible to heat up sufficient quantities of hot water on a range. But before water pipes were installed, rinsing was a problem. Few wells in Helsinki provided water in abundant enough quantities for it to be used to rinse laundry. There were only few streams and ponds in the city where laundry could be rinsed. Therefore, laundry typically had to be transported by means of handcarts or sleds to the nearest shore, to be rinsed in seawater. This was a natural choice because Helsinki is a coastal city situated on a peninsula surrounded on three sides by shallow bays of the Baltic Sea.

In the summertime, the rinsing of clothes at the seaside may even have been pleasant work, as women could chat on the warm jetty and enjoy the glittering waves, with the gulls and terns flying overhead. But after the short summer ended, this was no longer the case, as the following letter to the editor written by one eyewitness in April 1853 indicates:

> Servants here in the north, while performing the already-sufficiently heavy duty of washing clothes, are, during eight months of the year, exposed to the direct impacts of the harsh climate and not seldom compelled to complete washing at a hole in the ice in temperatures as cold as minus 20 or 30 degrees, while subjected to severe winds and snowstorms, protected merely by frozen-stiff clothes.[15]

During winters, the coldest place in the city was the frozen sea fronting the city, where the wind blew in off the icy Baltic without any obstacles. Wind dramatically increased the impact of the cold. An air temperature of −10°C in calm weather rapidly decreased to −30°C when the wind blew at a velocity of ten metres per second, a rather common hard wind on the open coastline, but not yet stormy. In such conditions, bare skin was in instant danger of freezing.[16] Naturally, washerwomen were aware of the risks winter brought, and cold was not an imminent danger if the women wore enough clothing and managed to keep it relatively dry. However, washing clothes first in a steaming hot sauna or washhouse and pulling dozens of kilograms of wet laundry to the shore on a handcart or sled meant that washerwomen were sweating before they even stepped on the ice, where the first thing they had to do was cut a rinsing hole in the ice with an axe or ice pick.[17]

Handling wet laundry on the ice, with seawater splashing up from the hole – not to mention the melting ice, slush, and snow – slowly but surely soaked the women's shoes, socks, and clothes. The women alternated between hot and cold sweats. The cloth screens that were occasionally hoisted to protect the washerwomen from light wind did not prevent the gales from blowing in off the glacial Gulf of Finland. Freezing cold water, wind, and hard work stiffened fingers and limbs. Naturally, if a woman lost a piece of clothing in the ice hole, she had to compensate her employer for it. Rinsing clothes and bedclothes with bare hands in freezing cold seawater through a hole in the ice was back-breaking work. Then the rinsed clothes had to be wrung out with stiff fingers to squeeze out the excess water before putting them back in the wooden tubs to be pulled in handcarts or sleds back to courtyards in town, where the clothes were hung to dry. Even though a household's entire laundry was washed only a few times during the wintertime, the rinsing took place at the seashore.[18] Washerwomen did not have to walk all the way to Siberia, like convicts in Imperial Russia did, to find Arctic conditions for labour. For women living and working in Helsinki, Siberia was as close as the nearest seashore.

Figure 9.1 This is the only known photograph showing laundresses at work in wintertime, when laundry had to be rinsed in holes that women hacked into the ice fronting the city. After the laundry had been rinsed by hand in the freezing water, it was wrung out, and the heavy loads were pushed by sled to be dried in the attics of residential buildings. Photo: G.A. Stoore, photo archives of the Museum of Central Finland.

Exposure to harsh weather was not purely a consequence of social order, but it aggravated social differences. According to one Russian physician, most diseases of working-class women were caused by laundry work in cold conditions.[19] For the women rinsing laundry on the ice, the Nordic climate was in the worst-case scenario a deadly serious problem. Hard work at the seaside increased the risk of accidents, diseases, and premature death not only for washerwomen but also for their children, who often accompanied their mothers.

The residents of Helsinki recognized the misery of washerwomen, and some solutions were proposed publicly in the 1850s. One person reported that in London as early as 1845 an association dedicated to the promotion of cleanliness among the poor had constructed special baths and a washing and drying house for free public use, and in France, steam engines and special washing apparatus were used to wash laundry. Another commentator mentioned that in the capital of Russia, several heated rinsing houses with walls and a roof were planned to be erected on the banks of the Neva River. There were even proposals to build laundrettes for washerwomen, with piped water, metal basins, and driers.[20] In short, already-tested technical solutions existed to help those people who suffered most from the cold Nordic climate.

Yet, the City of Helsinki did little to protect washerwomen in the mid-nineteenth century. An alert inhabitant noticed that, in 1860, the City of Helsinki had built a new wooden washing jetty by the seaside, but that it was open to the elements, with nothing to shield the exposed women. The observer commented the consequences of this somewhat fatalistically: "What diseases, what poverty, what misery is caused hereby; and all this, which in the end affects the whole municipality, could have been so easily prevented."[21] Women were at the mercy not only of the climate but of urban society as well. Poor washerwomen suffered from an archaic kind of "white slavery" on the snowy coast of the frozen Baltic Sea.

Short-Lived Hope for Washerwomen

Some members of the governing class of Helsinki had, however, noted the plight of washerwomen and started looking for new technical solutions to create a modern and efficient alternative to laborious home laundering. In March 1884, the first mechanized self-service laundrette opened in Helsinki. This laundrette was housed in a handsome, two-storey stone building near the centre of town. On the lower level, there was a dressing room for the washerwomen and lockable tubs where laundry could be left to soak overnight. The main room contained twenty-four "soap-saving" washing basins supplied with water from cold- and hot-water taps. The washing water came from the municipal water facility, which had been completed in 1876. The clothes could also be washed in lye steam. Clothes were rinsed in mid-room basins with seawater pumped from a nearby bay. After a steam-engine-powered spin, the laundry was easily transported by lift to the drying room upstairs, which was also equipped with steam mangles, an oven for heating up irons, ironing boards, and the holy of holies: a stove for the washerwomen's coffee pots.[22]

The self-service laundrette was designed by Helsinki's perhaps best-known architect, Theodor Höijer, working in cooperation with an engineer, K.M. Moring. Helsinki's largest machine shop built the machines. The laundrette company, Helsingfors Tvättinrättning, operating as a charitable association, was behind the initiative. The chairman of the board was the manufacturer F.M. Grönqvist, who had familiarized himself with comparable laundrettes in Paris, Berlin, and Vienna.[23]

The laundrette was marketed through an extensive advertising campaign. Mistresses could effortlessly schedule a laundry day for their maids by telephone. Use of the laundrette was free during the opening week, as an introductory offer. All spots were, indeed, taken during the first days the facility was open in March. A local Swedish-language newspaper warily asked, however, how packed the facility would be when it started charging.[24] The introductory offer was indeed followed by a drop in the number of customers, and the board of the laundrette company was forced to lower the fee from thirty pennies to twenty-five pennies an hour to attract a clientele.[25] The use of the laundrette continued to decrease in April with the onset of spring. In summer, it seemed as if the existence of the facility had been entirely forgotten. By autumn, the tone of news items was one of accusatory resignation: "It may take a long time, perhaps several decades, before we see the arrival of a new, comparable facility." The first modern laundrette in Helsinki rapidly went bankrupt and was closed.[26]

The success of laundrettes depended as much on their customers' social values and behaviour as on technology. The primary reason for the decrease in the use of the self-service laundrette was that, with the onset of spring, there was a return to the traditional and less expensive way of laundering: washing in a washhouse and rinsing at the seashore. "Old habits die hard," noted a local newspaper.[27] Still, the private laundrette had been built to meet the complaints of the city's womenfolk in particular. Homemakers had complained about the problems caused by the long winter, the harsh climate, and laundering in home conditions.[28] In spite of these complaints, the washing and rinsing of laundry were to incur as few expenses as possible. The community of Helsinki proved to be colder to the needs of the washerwomen than the wintry climate of the Gulf of Finland.

The washerwomen were thus forced to return to the exposed seaside. Contemporaries called attention to the icy tragedy being repeated on Helsinki's shores. This time, the health problems faced by washerwomen were not framed as restricted to the shores but rather social problems in the women's homes as well:

> Any doctor can attest to the misery and plight of the washerwomen, every citizen at the fringes of town can attest to how children go cold and hungry when their mother is ill, and the blame is society's. These workers are, however, among our most hard-working, and should be cared for.

The "human torment" of washerwomen freezing on the ice in the winter was now considered a shameful blight on the city. Letters were written to the editors of local newspapers, demanding washerwomen be remembered during a period when animal rights were already receiving significant attention.[29] However, even though there were literate washerwomen, not one wrote to the papers herself, nor did washerwomen have their own publishing channels. Like mute animals, the washerwomen were the targets of external pity.

When the private attempt at assisting the washerwomen failed, the municipal government at long last stepped in. The City of Helsinki began to build rinsing houses for public use in the 1880s and 1890s. The rinsing houses represented simpler and cheaper technology in comparison with the mechanized laundrette. The rinsing houses were floating wooden structures with a hole in the floor, through which the laundrywomen could rinse their laundry in the seawater. There was a lamp at the ceiling and a stove to the side, which could be used to generate some semblance of warmth. The rinsing houses did not, of course, warm the seawater, in which the laundry continued to be rinsed with bare hands. Still, at least the worst snowstorms and gusts of wind were blocked by the plank walls erected by the city. Most importantly, access was free to all. Despite their deficiencies, the rinsing houses meant a significant improvement in the working conditions of Helsinki's womenfolk. At the beginning of the 1900s, there were approximately ten municipal rinsing rooms on Helsinki's shores (Map 9.1).[30] The washerwomen's life sentence in Helsinki's own Siberia, the icy Baltic, finally came to an end – or rather, it seemed to come to an end.

From a Problem of Nature to One of the Environment

In April 1908, according to one Helsinki matron who wrote a letter to the editor, washing laundry more or less rolled along in well-equipped washhouses, where there was running water, lighting, and heat, "but in the end, the laundry had to be taken to one of the city's rinsing houses." The articles of clothing had to be rinsed along with dozens of others in the same icy holes, whose water was "polluted, dirty, and foul-smelling." The writer doubted that a better way of spreading disease had ever been invented and asked rhetorically who was responsible for maintaining hygiene and order in municipal rinsing

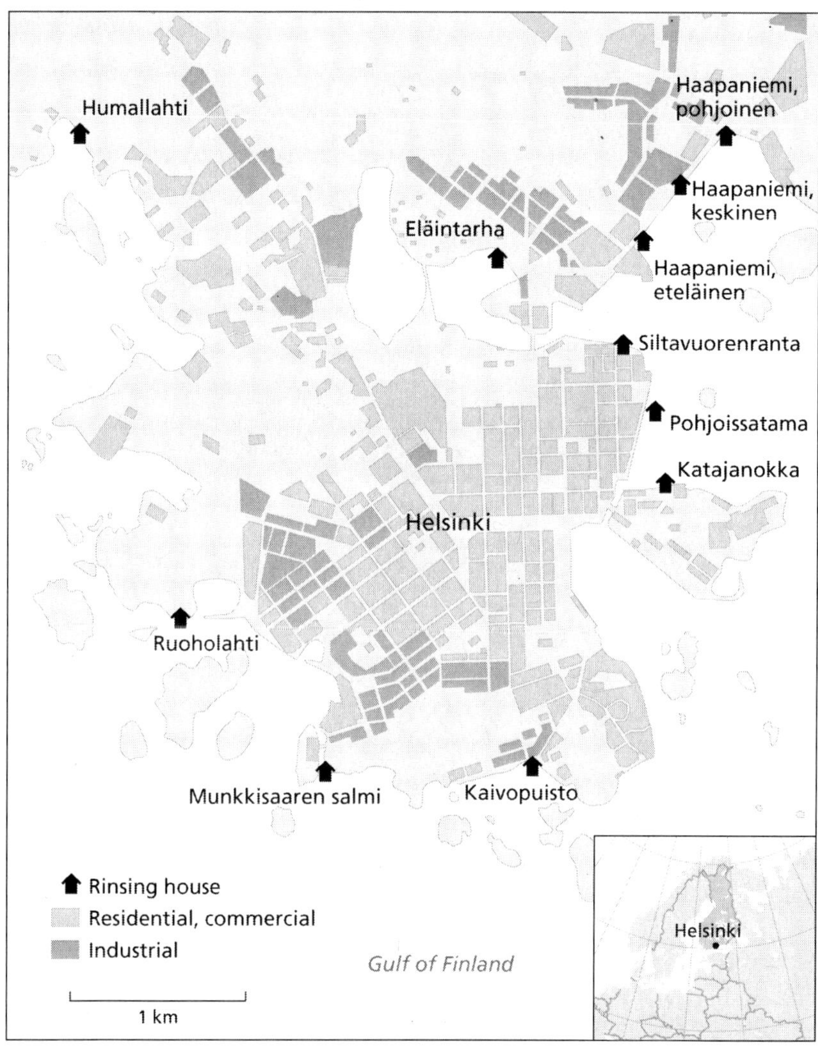

Map 9.1 The municipality of Helsinki placed eleven rinsing houses on the city's shoreline. The rinsing houses significantly improved the working conditions of Helsinki's women until the structures were closed in 1911 because of the pollution of seawater. The map is based on Helsingin kaupungin tilasto I. Terveyden ja sairaanhoito 2 (Helsinki, 1911) and photos of rinsing houses at the photo archives of the Helsinki City Museum.

houses.[31] It is not a surprise that the letter was written during the cold season. In terms of laundry rinsing, water pollution was a bigger problem in wintertime than in summertime. In summer, wastewater was relatively rapidly diluted by wind and waves. But in winter, wastewater, which was fresh water and therefore light, remained a thin yet concentrated layer between the ice and the salty seawater.[32]

The city took rapid corrective action. The inspector from the Board of Public Health declared that sewer water had spoiled the waters of two rinsing houses; these rinsing houses were closed down within a mere week of the matron's letter.[33] The City of Helsinki had, indeed, been forced to close rinsing jetties and houses earlier as well. In 1877, the city engineer decided to construct an unprotected rinsing jetty in a west Helsinki bay, where the seawater was considered clean. Twenty years later, the city health inspector noticed that the water inland from the breakwater contained so much bacteria that it could be compared in terms of consistency and smell to sewer water. The health inspector closed the rinsing jetty, even though the residents of the area used it "with discernible enthusiasm."[34] The residents were thus prepared to launder their clothes in polluted shoreline waters if no other practical alternatives existed.

Complaints about the dirtiness of seawater remained sporadic until the summer of 1908, when G.K. Bergman, a chemist and recent graduate of the University of Helsinki, studied the pollution of the city's sea areas, including swimming houses and rinsing houses. Bergman performed a test in a rinsing house located in the centre of town: first he boiled three regular handkerchiefs in lye, then he rinsed one of the handkerchiefs in the seawater from one of the rinsing houses in the centre of town, discovering 51,200 coli commune bacteria per cm^3. "How many bacteria infect a single sheet, let alone an entire load of laundry, when it is rinsed in one of our rinsing houses?" he asked. Bergman felt that the effects of rinsing on the cleanliness of laundry were "illusory" in Helsinki's coastal waters. Bergman's study, the results of which were published in every major newspaper in the city, relied on scientific authority to call attention to problems related to washing and rinsing laundry. Rinsing laundry thus became a significant societal problem because of the contamination of seawater, and in solving this issue, Helsinki clearly lagged behind its Nordic sister cities, Oslo, Copenhagen, and Stockholm.[35]

A follow-up study confirmed this assessment of the deplorable condition of the water at rinsing houses. In 1911, the municipal health laboratory took a series of samples from eleven rinsing rooms. Near the open sea, the bacterial content was only approximately 400 bacteria per cm^3. In sparsely inhabited areas, the water at the rinsing houses was also relatively clean; the average bacterial content was 16,700 per cm^3. The contents in more polluted rinsing houses were almost forty-fold in comparison: 602,000 bacteria per cm^3. In all rinsing houses, the water was yellow-brown, cloudy, and in several cases also foul-smelling.[36] Still, it must be noted that rinsing houses were not generally located on the filthiest shores. By means of zoning, the cleanest parts of the urban coastline were devoted to swimming and rinsing, whereas the most polluted areas were reserved for carpet-washing, boating, and industrial uses.

The city fathers were now forced to take action. In the winter of 1911, the Board of Health closed the public rinsing houses where the seawater was badly fouled. In the spring of 1911, nearly all of the city's rinsing houses were closed, and the use of many of the rinsing jetties was also forbidden. The board, which was led by men, did not, however, indicate any alternative locations for the rinsing of laundry. The laundry of the city's hundred thousand-plus inhabitants had, of course, to be washed and rinsed somewhere. The situation took a tragicomic turn. Now the washerwomen were forced to chop holes in the ice next to the locked rinsing rooms in order to rinse the city residents' laundry in the only feasible place, the sea.[37] So only a few decades after the rinsing houses finally became a reality, they were closed and the washerwomen had to return to their Siberia, to be wind-lashed on the ices of the Gulf of Finland.

Mistresses and Maids

The women of Helsinki did not wait around for the men to take action. In 1911, three Helsinki-based women's organizations demanded concrete action from the City of Helsinki to solve both the rinsing house closures and the polluted shorelines. This action marked a vigorous entree by women into environmental issues, perhaps the most vigorous in Finland to that point.

Behind the demands were three organizations: Kvinnoförbundet Hemmet (Women's Home Association), Husmoderförbundet (Homemakers' League), and Työläisnaisten liitto (League of Working Women). The members of Kvinnoförbundet Hemmet represented primarily the upper class of the capital city. In accordance with their agenda, the association strove to improve the moral, intellectual, and material resources of homes.[38] The Husmoderförbundet was an association for women from the Swedish-speaking cultural elite. Like Kvinnoförbundet Hemmet, the Husmoderförbundet met once a month, with the exception of a summer break. The association primarily discussed practical topics of relevance to homemakers, such as child rearing, the usefulness of the scouting movement, and ironing and cooking with electricity.[39] Most of the women from the middle and upper classes represented the traditional, well-educated Swedish-speaking elite. However, the organizing of working-class women also started from a bourgeois foundation. The

Helsingin Palvelijataryhdistys (Helsinki Female Domestic Servants' Association), established in 1898, was meant as a discussion forum for both mistresses and female servants. When founding member Miina Sillanpää, a former factory worker and maid, was selected as the chair a couple of years later, the association gradually began transforming into a labour union.[40] It also radicalized politically, joining the Työläisnaisten liitto and the Finnish Social Democratic Party. The women in the workers' organizations were for the most part poor, poorly educated Finnish-speaking girls from the countryside, many barely capable of reading and writing. The themes of their discussions were as distinct from those of the bourgeois women as coarse wool is from lace. The main topics covered in the organization's journal, entitled *Työläisnainen* (Working woman), were shorter working hours and wage increases.[41]

These women's organizations represented the different realities of the women of Helsinki. At the turn of the nineteenth century, there were numerous conflicts between women employees and women employers, particularly regarding the terms and conditions of employment. The political opposition between bourgeois and working-class women's organizations also rose to the fore in 1911. Because the domestic servants' working hours were undefined at the beginning of the 1900s, servants demanded two free evenings a week from the mistresses. The Husmoderförbundet and especially the Kvinnoförbundet Hemmet fiercely opposed the entire notion of a common free weeknight.[42] The Helsinki women's movement was, at the beginning of the 1900s, linguistically, culturally, socially, and politically deeply divided.

Reliance on Experts in the Environmental Policies of the Kvinnoförbundet Hemmet

On 20 March 1911, the Kvinnoförbundet Hemmet founded a committee to solve the rinsing house issue. By the next meeting, the members of the committee announced that they had contacted the gentlemen on the City Council and the Board of Health and demanded that the water problem at the rinsing houses be resolved. The answer given privately by the city fathers was that the city had already done all it could.[43]

Naturally, this answer did not satisfy the association. The members felt that the City of Helsinki needed to invest in solving the problem, for instance by laying the sewage pipes further out to sea, as had been done in other cities. Another option was that Helsinki would, like other cities, found a rinsing house where clothes would be rinsed with tap water in concrete basins. The City Council discussed the Kvinnoförbundet's proposal in June, deciding that it did not demand action. The women's association was tenacious, however, refusing to leave the matter there. In a new petition to the City Council, the association suggested modifying the sewage system in such a way that "the filth that now pollutes our shores should be reclaimed for agricultural purposes." The association referenced the system of irrigation fields in use in Berlin: sewage water was drained into and absorbed by fields near the city, where it was then possible to cultivate various products. By adopting a filtration system, "harmful pollution could even be put to use."[44]

The Kvinnoförbundet Hemmet also attempted to exert influence by calling in experts. The association invited, for instance, chemistry professor Ossian Aschan from the University of Helsinki to its meetings. Professor Aschan felt that it was inevitable that the city's growing population would continuously produce increasing amounts of waste. How that waste was to be handled was, however, a scientific, technical, and political issue. Aschan commented on the Kvinnoförbundet's proposal of establishing an irrigation field. According to the professor, the system worked in Berlin, but adapting this technology for use in Helsinki would nevertheless prove difficult because the price of land was so high that it would be difficult to acquire a large enough piece of land cheaply enough for the purpose. Nor was there enough sufficiently porous soil in the Helsinki vicinity. If wastewater were used to irrigate the clay beds surrounding the river running near the city, the source of the city's drinking water would be contaminated. What could be done to improve the situation? Aschan no longer considered prohibiting water closets viable, as their use was already widespread. The functioning of the existing biological wastewater treatment plants could still be improved by building more efficient facilities. As the best solution, Aschan proposed collecting the city's wastewater and conducting it through major sewers out far enough to open sea, where the currents would carry the waste away from Helsinki's shores.[45] Out of sight, out of mind.

A new problem for the women of Helsinki was, however, the fact that after the closure of the rinsing houses, many were forced to rinse their laundry in tap water, which had to be paid for. According to Aschan's calculations, an average load of laundry consumed approximately two to three cubic metres of water, in which case the costs would total seventy-five to ninety pennies. In his view, this sum was insignificant when weighed against the benefits. Aschan emphasized that under all circumstances the effort should be made to use solely tap water to rinse laundry. The only problem was that even though a municipal water system had been established in Helsinki in 1876, it did not generally aid washerwomen. The owners of residential buildings, for all practical purposes exclusively men, considered tap water too expensive to use for rinsing laundry. This is why laundry typically still had to be taken to the nearest shore for rinsing in seawater. Hence, there was a technical solution already in place, but men considered it too precious to be used.

The Kvinnoförbundet Hemmet thought it unfortunate that the Riesel-system in use in Berlin was not suitable for use in Helsinki because of the city's clay-heavy soil. The meeting proposed that information about the health risks of shore waters be broadly disseminated by brochure. The Kvinnoförbundet Hemmet came to the conclusion that the best way to get clean, fresh-smelling laundry was to pay a little for the use of tap water. It might even turn out that laundering would be slightly cheaper than before, since under the current system the householder had to pay to haul the laundry to the rinsing house and back.[46]

Reports from women's association meetings and Professor Aschan's presentation were published regularly in the city's main Swedish-language newspaper, a significant channel for influencing the opinion of the general public, half of whom were Swedish-speaking, as well as the city's decision makers, who were all Swedish-speaking. At the end of 1911, the city's Department of Public Works forwarded the Kvinnoförbundet Hemmet's presentation for renewing the sewage system to the Street Department, which was to prepare a proposal for measures to be taken. No actions were taken, and by the next year, Kvinnoförbundet Hemmet seems to have abandoned the matter.

Husmoderförbundet: Political Clout and Connections

At the same time that Kvinnoförbundet Hemmet was discussing the issue, the Husmoderförbundet proposed to the Helsinki City Council in April 1911 that the city's rinsing houses and jetties be moved to cleaner waters.[47] After its summer break, the association petitioned the city, urging it to speed up the construction of new rinsing houses. In addition, it demanded that rinsing basins be built in the washhouses of residential buildings. The association considered it of utmost importance that the price of the tap water used for rinsing be lowered as much as possible.[48]

In its response, the Public Works Department concurred with the Husmoderförbundet regarding the sensibility of the proposals, yet it questioned who was responsible for the execution of the measures. The board considered the proposal to build rinsing basins in the yards of private buildings as interfering with private property. Such measures exceeded the authority of the city. The Board of Public Works, however, let it be understood that the city could be held partially responsible for the pollution of the shores. It proposed building three rinsing houses in various locations around town with cement rinsing basins inside them and piped water.[49]

The discussion of the Husmoderförbundet's proposed solutions continued the following year. The City's *rahatoimikamari* (chamber of finances) indicated its concern that the prohibition on rinsing house use had caused significant dissatisfaction among the general public, particularly women's organizations. The chamber dissented only in terms of the number of public rinsing houses to be established and the fee charged to users. The majority felt that all three rinsing houses should be built at the same time and that the fee should be set at sixty pennies an hour per rinsing basin.[50]

The highest powers in the City of Helsinki took the Husmoderförbundet's petition seriously. The relative success of the homemakers' association might potentially be explained by the men who stood either behind or in front of the women. The chair, deputy chair, and secretary of the association apparently enjoyed good connections to the members of the City Council, which included their relatives or husbands.[51] The women were even perhaps able to compensate for a lack of formal political power by wielding informal but strong influence through familial ties.

The report by the chamber of finances was the last trace of the topic of rinsing houses in 1912, documenting that the three new rinsing houses with tapped water had been completed. Although the chosen technological solution did not directly address the problem of water pollution, as Kvinnoförbundet Hemmet's proposal would have, it did address the immediate issue of contaminated laundry. As a result of persistent activity on the part of women, at least a temporary solution had been found to this untenable problem.

The Environmental Perspective of the Women from Työläisnaisten Liitto

The fouling of Helsinki's sea areas was most concrete for the women who washed laundry for a living. No other people in Helsinki were in such constant, close contact with polluted seawater as the washerwomen, or, as they called themselves, *pesijättäret*, laundresses. The situation changed fundamentally when they organized at the end of the nineteenth century. The laundresses, who had previously been at the bottom of the societal hierarchy, transformed from objects of external pity to independent actors who took the future into their own rough but strong hands. After organizing, for the first time they also had their own trade union, female workers' newspaper, and representatives, through whom they could make their opinions known. The social distress of the washerwomen did, however, generate hurdles to activity. The profession was taxing and the pay poor.[52]

The laundresses acknowledged their weaknesses and recruited the assistance of the leader of Työläisnaisten liitto, Miina Sillanpää, who was a seasoned performer, a fine writer, bilingual in Finnish and Swedish, and in possession of political clout. Under the leadership of Miina Sillanpää, the working women demanded that the Helsinki City Council build modern washhouses, because the rinsing houses that had previously been at the shores were closed due to the foul-smelling and unhealthy seawater. The following extract from the meeting minutes of 12 September 1911 depicts the serious consequences that the closing of the rinsing rooms had for the entire working-class population, not just the laundresses:

> Particularly those who live on the west side of the city tell of flagrant negligence. The city has allowed rinsing locales to be closed due to the dirtiness of the water and now the washerwomen have to rinse their clothes from those bridges where carpets are washed. In even worse sludge, that is. There is no alternative, as in the first place most washhouses do not possess rinsing facilities, and the owners of many buildings have used the closure to their advantage, now extorting two or even three marks per load for rinsing. The working classes cannot afford this, especially large families who have to wash laundry often. The horror that this kind of rinsing inspires is proven by the fact that in many families, dishes must be dried in some manner other than wiping, due to fear of bacterial contamination from dishcloths that have remained dirty. All who spoke agree that the city is obliged to provide residents who pay substantial taxes to provide at the minimum facilities for washing and rinsing clothes, so that the working class doesn't need to starve in filth. In order to promote the matter, it was decided to turn to the City Council and demand that residents no longer be forced to rinse in even fouler water, that it begin procuring rinsing houses around town and that these be connected to the municipal water system, because due to deficient sewage facilities, any waste water pollutes the seawater in the vicinity of the city.[53]

In its petition, the executive board of Työläisnaisten liitto demanded that the city build a sufficient number of rinsing houses in working-class areas of town as well. As long as the seawater surrounding the city was polluted, tap water would need to be used in the rinsing houses. The old rinsing houses should be temporarily opened for city residents to use until the new ones were ready. In addition, the executive board listed the galling defects in the carpet-washing jetties in detail.[54] Thus, not only was the correction of major technical systems left at the doorstep of the city's masters, but the holes in the washing jetties needed to be nailed shut first thing.

The Board of Health immediately responded to the petition. According to the board, the rinsing houses were closed to prevent the spreading of pathogens and would remain closed. From a health viewpoint, it would, of course, be wonderful in the board's opinion if the City was able to build proper rinsing houses, but because they were very expensive, the board recommended the construction of rinsing basins in every washhouse at the expense of the renters or the property owner: in other words, a private solution was proposed to a municipal problem.[55]

The response of the city fathers did not satisfy the laundresses. The women felt that the Board of Health had adopted an arrogant stance of defending the irresponsible closure of the jetties without offering any alternative. Members wondered with what funds the working-class population was supposed to purchase rinsing basins for the buildings. The Työläisnaisten liitto decided to urge the City Council to coldly forget the idle statements by the Board of Health and attend to the wishes of the women's chapter in this matter.[56] The promised improvements did not appear, however, and the dirty laundry continued to be aired in the newspapers in 1912. Laundresses expressed frustration that building owners extorted two or three marks for rinsing a small load of laundry in tap water. The price was considered impossibly high for the poor, whose wallets were painfully sensitive to the tiniest increase in the costs of living. Miina Sillanpää wrote an article in *Työläisnainen* in which she sharply criticized the board for failing to understand that, through its health policies, it was causing a substantial increase in the living costs of the poor.[57]

At the monthly meeting of the Työläisnaisten liitto in April 1912, there was discussion about the chamber of finance's plan for new washhouses and the fees for using them. Particular indignation was caused by the city's intent to "begin extorting" sixty pennies an hour for use of the rinsing houses. The final comment of the laundresses on the plans of the City of Helsinki was unforgivingly blunt: "If it [is] not reasonable to establish free rinsing houses in the city, then the poor will have to stop washing clothes until they split with filth, no matter how foul they smell and how much disease they spread."[58]

A Touch of Frost?

In this study I tried to understand the impact of industrialization on laundry work, the social context of laundry work, and the urban environment where it took place in Helsinki. The starting point of the study was the fact that the industrializing city was divided in two according to gender. The city's bureaucracy and political organs were predominantly male-dominated organizations, and in households, economic power and decision making were generally the province of husbands. The city's gender structures were also laid bare in the laundry issue. The water and sewage system was a large-scale technical system planned, built, and managed by the elite of the industrializing society: by educated, wealthy, Swedish-speaking men. Yet, the man-run technical system was in everyday circumstances used most frequently by the lowest caste of society, generally Finnish-speaking, nearly illiterate women who were often recent transplants from the countryside. While the city fathers strode the boulevards of the city in galoshes with dry feet, poor women had to work in cold, dirty coastal waters almost barefoot.

Although women who were experts in laundry work took care of the work itself in practice, it was uninformed and occasionally ignorant men who were responsible for planning, building, and maintaining the related technical facilities. Therefore, laundry work is another good example of the gender gap and its consequences – that is, of a situation where best available solutions are not adopted because of a deep, gender-based division of labour and related lack of knowledge and political power. Hence, the gender gap, the overrepresentation of either sex in an issue that applies to both sexes in general,[59] explains to a high degree the slow development of societies in certain fields.

Yet, when in their ignorance the men prevented the rinsing of laundry in the new rinsing houses, the women took action, though their approach was not monolithic. Despite a common problem, when taking action the women of the capital city were divided into different social groups, with different reasons for and opportunities to influence municipal decision making. The Kvinnoförbundet Hemmet represented the elite of society. It had abundant resources and opportunities to influence decision making, but the will of the association to dedicate themselves to the matter appears to have ended earlier than for the other women's organizations. The more middle-class Husmoderförbundet enjoyed, in addition to an educated leadership with international experience, access to the city's leading Swedish-language newspaper, as well as good political connections to the bourgeois City Council. The will of the more middle-class Husmoderförbundet to influence laundry arrangements was also stronger than that of the more upper-class Kvinnoförbundet Hemmet, for whose members laundering was in practice a rather distant matter in any case. The laundresses themselves meanwhile also aired their grievances, but with little political clout, their complaints were unheeded.

So far, most studies that have linked women and the environment stress women's involvement as an extension of their roles as mothers and protectors of the home and community. Women tended to not challenge the existing power hierarchies because they had little access to formal, male-dominated public power. Instead, middle- and upper-class women mainly aimed to realize reforms by means of educating members mentally and morally, creating public opinion, and securing better life conditions.[60] This case study supports these conclusions on a general level, with the exception that, in this Nordic city, women from the working class also proved active in environmental issues, in this case marine pollution. But what then explains the fact that in the Nordic city, working-class women were also active in environmental issues?

The primary reason for the laundresses' strong desire to influence the cleaning of the fouled seawater was that they were forced to be in direct contact with it on an almost daily basis, though this was hardly anything new for them. Another significant reason was that the laundresses of Helsinki, who had traditionally been at the bottom of the hierarchy of urban society, had independently organized into a trade union. As a result of political activity and cooperation with other working-class women and their organizations, an entirely new sense of self-worth and solidarity was born, which was imperative for the societal emergence of the laundresses. An additional critical factor was the parliamentary election held in Finland in 1907, in which Finnish women were the first in the world to receive complete political rights – in other words, the right to vote and to stand as nominees in national elections. A total of nineteen women were elected to Parliament then, of which almost half represented the working class. This event, in addition to having enormous symbolic significance for these women, had large practical significance, as the laundresses now had an ally at the pinnacle of power, a female Member of Parliament with a working-class background. Miina Sillanpää was the leader the women of the Finnish Social Democratic Party and proved to be an exceptionally energetic and skilled practitioner of municipal politics as well. Approximately a decade later, she was named Finland's first female minister. Washerwomen had a powerful ally.

A major motivation for all groups of women to act was that laundry was an important part of the identity of women. During the period under examination, all women considered laundry work to be their responsibility. Women from all social classes were concerned about the polluting of the seawater and the problems it caused for laundering. They were not, however, of one mind as to how to resolve the situation. The fundamental difference between the reactions of working-class and bourgeois women to solving the problem of fouled seawater was economic. Once the seawater that surrounded the city had been polluted, the bourgeois women's associations were prepared to pay for rinsing laundry with clean tap water. The Työläisnaisten liitto was not. Although the cold, polluted seawater was a dangerous element for working women, it was also a free resource for them, one that had significance for the entire livelihood of working-class families.

Gender united but class divided women. In the Nordic countries, women – from the working class to the middle and upper classes – organized particularly actively. Primarily because of conflicts related to terms and conditions of employment, the women of Helsinki were already deeply divided by the late 1800s. The Social Democratic, bourgeois, and conservative women's organizations simply could not present a united front, even on behalf of rationalizing laundering and protecting the waters. Historian Maureen Flanagan has concluded that the division of women often helped men to ignore women on a variety of issues.[61] But would have a united front been of much use anyway in Helsinki? It seems that the deep economic, social, and cultural division and consequent political competition between different female organizations probably sharpened and sustained their activities individually and collectively. In this volume, Marionne Cronin has incorporated exploration of the North with concepts such as voluntarism, heroism, masculinity, and national identity. In the case of women, the conclusions are almost the opposite; facing the hard northern conditions was rather incorporated with coercion, anti-heroism, femininity, and group identity.

Did the activities of the three women's organizations examined here have any impact? The two-year conflict certainly accelerated the introduction of new urban and household technologies – that is, piped rinsing water into households, the construction of washhouses equipped with water pipes, and also the adoption of washing machines in Helsinki. Rinsing of clothes and bedclothes in the seaside thereby gradually dwindled during the first decades of the twentieth century. But the activity of women had a considerable impact on water protection in Helsinki as well. From the time of the campaigns conducted by the women's organizations, the City of Helsinki established a special committee to plan improvements to municipal water protection. As a result of the scientific and technical work conducted by the committee, the Helsinki City Council decided in the late 1920s to initiate an ambitious investment program to build activated sludge plants in order to purify all wastewater generated by the city. The Second World War somewhat delayed realization of the plan, but finally as all municipal and industrial wastewater was treated, the coastal waters around the City of Helsinki gradually became cleaner. Some years ago, Helsinki was chosen by a quality-of-living ranking as the cleanest capital in the European Union and the third cleanest in the world.

Integration of social history and environmental history provides a powerful methodological approach to study how power differentials in society were also felt in environmental ways. In practice, this integrated approach addresses the need for comparative studies of social and environmental dual exposure. Social exposure signifies the different probability of different social groups to face risks such as hunger, undernourishment, poverty, disease, or various forms of marginalization. Environmental exposure for its part signifies the disparate probability of different social groups to be exposed to extreme temperatures, humidity, windiness, luminosity, other related natural risks, or man-made pollution and other similar hazards. For Helsinki's washerwomen, dual socio-environmental exposure leads easily to a paradoxical situation where people who have the least socio-economic resources are forced to face the most extreme environmental conditions as well. As a result of this negative trickle-down effect, environmental exposure exacerbates social exposure and vice versa, creating a vicious socio-environmental circle that may further aggravate existing inequalities in a given society. In conclusion, it is suggested that concepts of socio-environmental exposure and environmental justice are fruitful starting points for future studies of the North as well.

NOTES

1 "Veckans krönika," *Papperslyktan*, 21 January 1861.
2 "Om väderleken I," *Finlands Allmänna Tidning*, 19 June 1854; for monthly minimum temperatures in Helsinki since 1844, see http://www.helsinginymparistotilasto.fi/.
3 "Tvätthus,"*Finlands Allmänna Tidning* (hereafter *FAT*), 26 March 1860.
4 Carolyn Merchant, ed., *Women and Environmental History*, special issue, *Environmental History Review*, Spring 1984: 1-112.
5 Elizabeth D. Blum, "Linking American Women's History and Environmental History: A Preliminary Historiography," http://www.h-net.org/~environ/.

6 See, for example, Suellen Hoy, "'Municipal Housekeeping': The Role of Women in Improving Urban Sanitation Practices, 1880-1917," in *Pollution and Reform in American Cities, 1870-1930*, ed. Martin Melosi (Austin: University of Texas Press, 1980), 173-98; Harold L. Platt, "Invisible Gases: Smoke, Gender, and the Redefinition of Environmental Policy in Chicago, 1900-1920," *Planning Perspectives* 10 (1995): 67-97; Angela Gugliotta, "Class, Gender, and Coal Smoke: Gender Ideology and Environmental Injustice in Pittsburgh, 1868-1914," *Environmental History* 5 (2000): 165-93; Maureen Flanagan, *Seeing with Their Hearts: Chicago Women and the Vision of a Good City, 1877-1933* (Princeton, NJ: Princeton University Press, 2002).

7 Arwen Pelmer Mohun, "Laundrymen Construct Their World: Gender and the Transformation of a Domestic Task to an Industrial Process," *Technology and Culture* 38 (1997): 97.

8 For a seminal study on domestic laundry work in Britain, see Patricia Malcolmson, *English Laundresses: A Social History, 1850-1930* (Urbana: University of Illinois Press, 1986); for the cited text, see Malcolmson, *English Laundresses*, 5; for information on laundry work in Germany, see Karin Hausen, "Grosse Wäsche: Technischer Fortschritt und sozialer Wandel in Deutschland vom 18. bis ins 20. Jahrhundert," *Geschichte und Gesellschaft* 13, 3 (1987): 273-303; for Sweden, see Gunnel Hazelius-Berg, "Tvätt," *Fataburen: Nordiska museets och Skansens årsbok*, 1970: 115-30; for laundry work in the French countryside, see Yvonne Verdier, *Façons de dire, façons de faire: La laveuse, la couturière, la cuisinière* (Paris: Gallimard, 1976).

9 For an overview, see Rudolf Brázdil et al., "Historical Climatology in Europe: The State of the Art," *Climatic Change* 70, 3 (2005), 363-430; H.H. Lamb, *Climate, History and the Modern World* (London: Methuen, 1982); for a notable exception, see Brian Fagan, *The Little Ice Age: How Climate Made History, 1300-1800* (New York: Basic Books, 2000); for Finland, see Antti Häkkinen et al., eds., *Kun halla nälän tuskan toi: Miten suomalaiset kokivat 1860-luvun nälkävuodet* (Porvoo, Finland: Werner Söderström, 1991).

10 See, for example, Stephen J. Pyne, *The Ice: A Journey to Antarctica* (Iowa City: University of Iowa Press, 1986); Bravo and Sörlin, *Narrating the Arctic*; Denis Cosgrove and Veronica della Dora, eds., *High Places: Cultural Geographies of Mountains, Ice and Science* (London: I.B. Tauris, 2009).

11 Arne Dufwa and Mats Pehrson, *Snöröjning, renhållning, återvinning* (Stockholm: Stockholms stad, 1989); Bernard Mergen, *Snow in America* (Washington, DC: Smithsonian Institution Press, 1979); Sven Lilja, "The Atmosphere of the City: Climate in the History of Stockholm," in *Living Cities: An Anthology in Urban Environmental History*, ed. Mattias Legnér and Sven Lilja (Stockholm: Formas, 2010), 220-57; Simo Laakkonen, "Jäätyneet kaupungit," *Historiallinen Aikakauskirja* 3 (1997): 217-26.

12 See Stephen Mosley, "Common Ground: Integrating Social and Environmental History," *Journal of Social History* 39, 3 (2006): 915-33.

13 For a contemporary discussion, see, for example, Paul Norton, "A Critique of Generative Class Theories of Environmentalism and of the Labour-Environmentalist Relationship," *Environmental Politics* 12, 4 (2003): 96-119.

14 Nina E. Lerman, Arwen Palmer Mohun, and Ruth Oldenziel, "The Shoulders We Stand on and the View from Here: Historiography and Directions for Research," in *Gender Analysis and the History of Technology*, special issue, *Technology and Culture* 38, 1 (1997): 15.

15 "Allmänna inrättningar för bad och klädtvätt, II," *FAT*, 9 April 1853.

16 Finnish Metereological Institute, "Ilmatieteen laitos," "Sää ja ilmasto," and "Tuulen nopeus," http://www.fmi.fi/saa/havainto_12.html.

17 This description of laundry work is also based on my experiences as a young boy helping my mother, Gunborg Laakkonen, rinse clothes on the ice when we were living on the coastline, where we had no access to piped water.

18 Simo Laakkonen, "Itämeren tyttären likaiset helmat," in *Nokea ja pilvenhattaroita: Helsinkiläisten ympäristö 1900-luvun vaihteessa*, ed. Simo Laakkonen, Sari Laurila, and Marjatta Rahikainen (Helsinki: Helsingin kaupunginmuseo, 1999), 109-38.

19 "Tvätthus," *FAT*, 26 March 1860.

20 "Sundhetsvård," *FAT*, 18 October 1852; "Allmänna inrättningar för bad och klädtvätt, I and II," *FAT*, 8 and 9 April 1853.

21 "Vanlig förfrågan,"*FAT*, 27 April 1860.

22 "Helsingfors tvättinrättningen," *Hufvudstadsbladet*, 6 March 1884; "Helsingfors tvättinrättning," *Helsingfors Dagblad*, 7 March 1884.

23 "Helsingfors tvättinrättningen," *Hufvudstadsbladet*, 6 March 1884; "Torgberättelse," *Hufvudstadsbladet*, 4 March 1882.

24 "Helsingfors tvättinrättning," *Hufvudstadsbladet*, 15, 16, and 25 March 1884, and 28 June 1884.

25 Penni are the divisions of the Finnish mark: 100 penni = 1 mark. The average daily wage of female agricultural workers (who paid their own meals) was 1.39 Finnish marks in 1910. This indicates rather well the wages in cities too, even though there are no statistics available for female urban workers' average daily wages. *The Economic History of Finland*, 3: *Historical Statistics* (Helsinki: Kustannusyhtiö Tammi, 1983), table 13.1. Daily wages in manual day work.

26 "Helsingfors tvättinrättningen," *Hufvudstadsbladet*, 25 March 1884; "Vid tvättinrättningsbolag ...," *Hufvudstadsbladet*, 1 April 1884; "Twättinrättningen," *Hufvudstadsbladet*, 27 April 1884.

27 Husmodern, "Till Red: Af Nya Pressen," *Nya Pressen*, no. 69, 12 March 1885; "Helsingfors twättinrättningen," *Hufvudstadsbladet*, 31 October 1884.

28 "Helsingfors twättinrättningen,"*Hufvudstadsbladet*, 31 October 1884; "Twättinrättningen," *Hufvudstadsbladet*, 15 April 1884; "Om tvättinrättningar och tvättapparater," *FAT*, 29 October 1870.

29 Nimim. Ihmisrääkkäyksen vihaaja, "Korjattava puute," *Uusi Suometar*, 13 January 1881; "Torgberättelse," *Hufvudstadsbladet*, 2 March 1882; "Klapphus åt tvätterskorna!" letter to editor, *Nya Pressen*, 3 December 1886.

30 "Oivallinen laitos..., En god inrättning," *Helsingfors Tidning*, 19 November 1864.

31 Nimim. -a., "Stadens sköljhus," *Hufvudstadsbladet*, 11 April 1908; see also "Tölö viks vatten, T.F. Polismästaren ingriper," *Hufvudstadsbladet*, 28 August 1908.

32 The committee charged by the City Council to put municipal sewage in order, 1916-19. Published documents of the Helsinki City Council, 4/1924, Appendix 2, report on chemical studies.
33 "Stängda sköljhus," *Hufvudstadsbladet*, 11 April 1908.
34 HKA, Public Works Department, Hans Homén's undated response to the proposal by the chamber of finances, 28 September 1877; to Magistraten i Helsingfors, report of health inspector Wilhelm Sucksdorff to the magistrate, 2 August 1893.
35 G.K. Bergman, *Tutkimuksia laskuveden vaikutuksesta vesiin kaupungin ympärillä kesällä 1908: Helsingin kaupungin terveydenhoitolautakunnan vuosikertomus, 1907-1908* (Helsinki: Helsingin kaupunki, 1908), 32; Keskuslaboratorio Archives, G.K. Bergman Collection, received letters 1908-20, Stockholms hälsovårdsnämd, response of Klas Sondén, 14 September 1908.
36 Statistics of the City of Helsinki I, Terveyden- ja sairaanhoito 2, 1911, 66-70.
37 *Hufvudstadsbladet*, 4 April 1911, 5.
38 *Hufvudstadsbladet*, 1 January-31 December 1911; "Toimintaohjelma," *Hufvudstadsbladet*, 25 March 1911, 10.
39 Untitled, *Hufvudstadsbladet*, 7 March 1911, 6.
40 Irma Sulkunen, *Naisen kutsumus: Miina Sillanpää ja sukupuolten maailmojen erkaantuminen*. (Helsinki: Hanki ja jää, 1989), 16-18; Martta Salmela-Järvinen, *Miina Sillanpää: legenda jo eläessään* (Porvoo, Finland: Werner Söderström, 1973), 7.
41 *Palvelijatar*, 23 January 1908.
42 *Työläisnainen*, 19 December 1912, 390-91; *Palvelijatar*, 23 January 1908; Sulkunen, *Naisen kutsumus*, 14, 86-87.
43 "Det orena vattnet i våra hamnar," *Hufvudstadsbladet*, 30 April 1911.
44 This and the preceding quotation are from "Kvinnoförbundet Hemmet," *Hufvudstadsbladet*, 29 October 1911 and 5 November 1911; *Hufvudstadsbladet*, 3 December 1911, 11.
45 "Om vattnets föroreningar i våra hamnar, Föredrag av prof. Ossian Aschan," *Hufvudstadsbladet*, 12 December 1911.
46 Kvinnoförbundet Hemmet, "Förorening af vattnet i hamnarna,"*Hufvudstadsbladet*, 14 December 1911.
47 *Hufvudstadsbladet*, 4, 5, and 23 April 1911, 11.
48 *Hufvudstadsbladet*, 22 October 1911, 9.
49 HKA, Public Works Department, Kirjetoisteet 1911, YTH letter no 353; *Hufvudstadsbladet*, 13 December 1911, 7.
50 *Hufvudstadsbladet*, 18 January 1911, 7; 9 March 1911, 6; and 11 March 1912, 2.
51 *Kertomus Helsingin kaupungin kunnallishallinnosta*, 1911, list of City Council members.
52 *Työläisnainen*, 20 May 1909, 149.
53 TA (Archives of the Workers' Association), minutes of the Helsinki Workers' Association (HWA) women's chapter monthly meeting, 12 September 1911. Undated newspaper clipping appended.
54 "Sköljhusen i staden," *Hufvudstadsbladet*, 5 October 1911.
55 "Stängningen af stadens sköljhus," *Hufvudstadsbladet*, 30 November 1911.
56 TA, minutes of the HWA women's chapter monthly meeting, 12 December 1911.
57 *Työläisnainen*, 22 February 1912, 63.

58 *Työläisnainen*, 12 May 1912, 142.
59 For another, more economic definition of the gender gap, see Claudia Golding, *Understanding the Gender Gap: An Economic History of American Wome*n (New York and Oxford: Oxford University Press, 1990), ix.
60 Blum, "Linking American Women's History and Environmental History."
61 Flanagan, *Seeing with Their Hearts*, 162.

PART 4

IMAGINING THE NORTH

North Takes Place in Dawson City, Yukon, Canada

LISA COOKE

Dawson City, Yukon, Canada, sits at the confluence of the Yukon and Klondike Rivers, approximately 2,900 kilometres north of Vancouver, British Columbia; 100 kilometres east of the Alaska-Yukon border; and 240 kilometres south of the Arctic Circle (Map 10.1). Dawson City's geographical remoteness, its northness, is central to its emergence on the map as a point of international significance. As noted throughout this volume, northern environments are not discovered at the peripheral edges of the circumpolar globe. They are created by the constantly shifting relationships between humans and nature in those spaces. By way of these complicated sets of relations between people, technologies, weather, seasons, animals, economics, and politics, northern environments are made. This chapter is about the making and remaking of Dawson City through these very networks.

I come at this discussion from the perspective of a cultural anthropologist interested in places. I think about places never as fixed, static sites of culture but rather always as emergent cultural processes.[1] Turning to cultural geographer Bruce Braun for guidance, I conceptualize places as *taking place* as events. Braun writes, "Places *take place*, they *happen*, they cannot exist apart from the flows that constitute them, that cull them into existence" (emphasis in original).[2] What this perspective offers is a means of thinking through the complexities of the ways that meanings, values, interests, and power relations become embedded in places and how places are

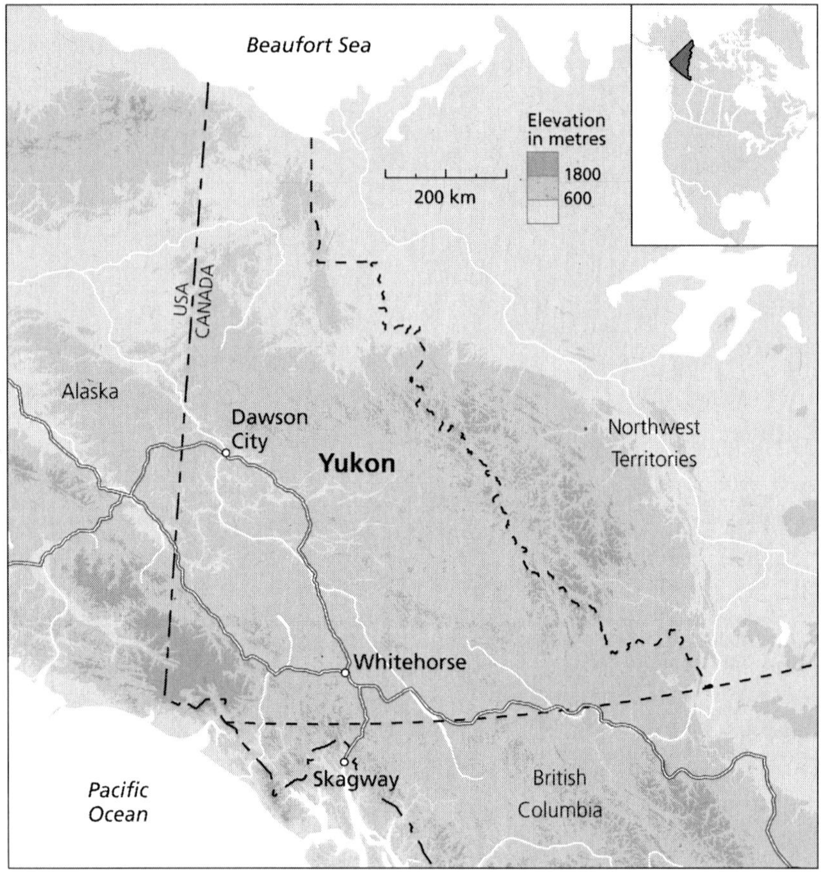

Map 10.1 Yukon Territory and the location of Dawson City

made in response to these forces. From here we must consider not just what places *are* but what do they *do*.

My anthropological and theoretical leanings make me well practised at thinking about the cultural complexities of place making. Traditionally, I have paid close attention to the bundles of cultural values and their associated power relations that come together and emerge into view as a place. Although I have nodded to places as being historically made, I have always paid more attention to the culturally produced lore about historical events than the events themselves. Fun as this might be to do, I see now (with much thanks to my generous colleagues in history) that what this offers is a somewhat one-dimensional view of things. I have been challenged by these colleagues to attend to time. I have been challenged further by my colleagues in environmental history to attend to time *and* space and to situate my analysis of the cultural processes through which places emerge, happen, firmly in both. Thank you for that. By thinking about places occurring in the compression between time and space, the making and subsequent remaking of Dawson City presented here becomes much more multi-faceted. In what follows I invite readers to join me as we journey north through the layers of meaning, politics, and history that constitute the very notion of North in Canada to Dawson City, Yukon. Once there, we'll pause on the banks of the Klondike River and examine the ways that memory, technology, and politics collaborate in the making of this place as a specifically northern environment. Gazing out across time and space, we will consider this northern place both for how it came to be *and* for what it does.

Dawson City: "The Heart of the Klondike Gold Rush"

Dawson City is coined "The Heart of the Klondike Gold Rush" for good reason.[3] It is. Neither Dawson City nor Yukon existed before gold was discovered in the area in 1896. To be sure, there were people and places and things happening in the region, but both Dawson City and Yukon Territory as marked locations on the historical and geographic maps of Canada and the world came into being as a result of the discovery of gold and the subsequent rush for more of it that followed.

The story of the Klondike Gold Rush goes something like this. On 16 August 1896, someone found a large nugget of gold in Rabbit Creek (later renamed Bonanza Creek) not far from what is now Dawson City in what is now Canada's Yukon Territory. Who is credited with having discovered the first nugget of gold in the Klondike depends largely on who is telling the story. In many American accounts, George Carmack of the United States is the originary protagonist. This makes it easier to tell the story of the Klondike Gold Rush as an American gold rush tale that follows the trajectory of gold rushes in California in the 1840s and served as the gateway to future rushes in Alaska in the 1900s.[4]

Canadian versions of the story, however, privilege tellings that insist that Canadian Robert Henderson first mined gold in the Klondike and tipped off Carmack. Indigenous historians suggest that it was actually First Nations colleagues of Carmack, Skookum Jim Mason and Dawson (Tagish) Charlie, who found gold while Carmack was asleep.[5] Whoever started it, it started, and by spring of the following year, as word of the strike was reaching the outside world, all of the creeks in the region had been claimed. That mattered little to those lured north by potential fortune. The global demand for gold that had fuelled a series of preceding rushes in California and British Columbia held strong, and an estimated thirty thousand people made the trek north to Dawson City in search of riches. The region came into being as the Klondike, and Dawson City its "heart."[6]

As people, goods, and capital moved north, networks of mobility expanded. Cities at entry points for the route to the Klondike – Seattle, Vancouver, and Edmonton – grew enormously as a result.[7] Boomtowns emerged on the Alaskan coast at Skagway and Dyea to serve as staging grounds for the arduous inland trek to the Klondike.[8] People travelled by railway to port cities, steamship up the coast, and by foot, boat, raft, and much of their own labour from there.[9] Environmental historian Kathryn Morse offers a richly textured account of the range of technologies of mobility employed as people moved themselves and their cargo north into the Klondike.[10]

In January 1897, Dawson City was officialized as a town site, and within a few short years transformed from a collection of tents to being the largest city north of San Francisco and west of Winnipeg. It was not just people moving thousands of kilometres into the Klondike. With them they brought goods, services, capital, and political aspirations. In this remote pocket of North America a particular kind of northern environment emerged. Rooted in a vision of a resource-rich hinterland free for the taking, the North came into view as a terrain upon which national ambitions could be inscribed – wilderness discovered, tamed, conquered, and civilized. The establishment of an industrial city in the country's northern reaches fed Canadian national narratives of nation building. Dawson City was not a rustic mining town on the fringe. Coined "Paris of the North," it was the first city in western Canada to have electricity. In a few short years, it emerged as a symbol of national northern possibilities. In a corresponding political assertion of sovereignty, the Yukon was established as a distinct territory of Canada in 1898, and Dawson City named its capital.[11]

Dawson City may not have existed before 1897, but people had inhabited the area for generations. The site that is now known as Dawson City is located in the traditional territories of the Tr'ondëk Hwëch'in. Across the water from the site where Dawson City came to be sits the Tr'ondëk Hwëch'in settlement at Tr'ochëk. As Dawson grew, the Tr'ondëk Hwëch'in were pushed up the river to a new settlement site at Moosehide, where they would remain for fifty years. In their place, Tr'ochëk became known as Klondike City, or Lousetown – the place across the river designed to contain the morally questionable elements of boomtown life.[12]

So in addition to being a tale of movement, mobility, mining, and the making of this region called Klondike into a northern resource-rich environment, the story of Dawson City and the Klondike Gold Rush is also one of frantic, rapid, and enduring displacement. Thousands of people arrived. Land was staked and claimed, and then bought and sold. Tr'ondëk Hwëch'in territory was transformed into property, a discursive gesture that denied indigenous relationships with the land and indigenous title and rights – a discursive gesture that is central in the making of this northern environment as a colonized place.

The story of the Klondike Gold Rush is the story of Canadian colonial expansion. In this volume, Julia Lajus discusses the Soviet quest to colonize the northern reaches of western Russian territories. In its efforts, the Soviet Union looked to Canada's colonial successes for guidance. What is most notable about this for our purposes is that the Soviets did not shy away from the intent of the project. Colonization was the stated goal, Canada the role model. In Canada, we frame the story slightly differently. The story of the Klondike Gold Rush is transcribed into historiography as a heroic pioneering tale of northern expansion and Canadian nation building. That Canadian national building was, and is, a colonial project sits uncomfortably at the back of our collective throat. It's the footnote that we would rather not reference. Instead, we celebrate those things done in the name of Canada building, turning our gaze away from the ghosts that haunt the margins of the project.

This book is concerned with the various technologies that bring new northern environments into being. Technologies shape and change the ways that environments are perceived, used, and valued. As seen in the chapters presented here, various technologies have been central in the production and making of northern environments. The airplane offered a new way of gazing, and thus perceiving, northern spaces. Railways and roads inscribe yet another structure of access and visibility. Driving into Dawson City today from the south, one cannot help but notice impacts of the technologies of mining and mobility on the land. Mounds of dredged earth zigzagging their way across the valley remind us that a lot of gold was mined from the creeks and areas around Dawson City. To facilitate this extraction, a city was built. People came. Trees were cut down. The landscape around Dawson City is littered with the traces left by these incredible efforts. These traces form an archive of the material impacts of the Klondike Gold Rush in the making of this environment. Contemporary travel north to Dawson City has many of us following these same routes.[13]

Environments are made by more than just material and mechanical technologies, however. The complex relationships between humans and nature out of which environments emerge are imaginative and discursive too. In the case of the making of Dawson City as a distinctly northern and national place, national-cultural narratives were, and are, deployed. Narrating a space into importance narrates it into view – claims it, shapes it. Political and economic interests underscored the efforts of the Klondike Gold Rush at the time. Colonial assumptions about land "discovered" being free for the taking justified its taking.

Over a century after the gold rush petered out, political and economic interests continue to underscore how Dawson City is made as a specifically northern and national environment. This is where material and discursive technologies join forces. The process of translating bits and pieces of past events into something we can understand as history is a narrative one. The Klondike Gold Rush is narrated into significance in the present as emblematic of dominant narratives of Canadian nation building. Dawson City becomes symbolic of a national Canadian pioneering dream and national mythology of the North as the site of resources and riches. Memory and politics collaborate, and once they take hold, these dominant national-cultural narratives are materialized on the ground through the production of official state-sponsored national heritage. Deploying the production of officialized national heritage is a key technology in the making of Dawson City that draws on the past to serve the needs of the present.

Material and discursive technologies do not operate in isolation from each other. Quite the opposite, as Bruce Braun suggests: "The discursive and the material ... implode into knots of extraordinary density."[14] Political and economic needs of the present continue to infuse the North with national aspirations and interests. The North, in Canadian national-cultural imaginaries, plays double-duty as simultaneously emblematic of a great national wild naturescape and a rich repository of resources to fuel the country's economic growth.[15] On the ground, Dawson City can be made in the image of both and travel to them a way to connect people to the story.

The same networks of mobility established during the Klondike Gold Rush are now used to transport visitors to the region. The economic base of the community is transferred from gold mining to commemorative archives of gold mining. The material traces left by the Klondike Gold Rush become sites of pilgrimage made meaningful in collective memories because of their symbolic capital as both emblematic of the North's resource-rich potential and Canadian nation building. Sites are restored. Plaques erected marking significance. To get people to them, more roads are built, parking lots created. Tourists need other things too – places to eat, sleep, and shop. Out of the discursive project of official state-sponsored productions of national heritage, new technologies meet old ones as they work to create a new kind of northern environment.

This knot implodes further when we recall that the story of the Klondike Gold Rush is also the story of colonial conquest and displacement. Commemorating and celebrating the Klondike Gold Rush as an emblematic event in Canadian nation building also means celebrating those things done in the name of that project. And if we are to consider the discursive processes of state-sponsored heritage production technologies of place making, then we must also consider them technologies in the work of colonial conquest. By way of the production of officialized national-cultural heritage in Dawson City, this northern environment is continuously remade as a colonial and colonizing place.

On the Banks of the Klondike River

On the banks of the Klondike River, one warm July afternoon I met Janice.[16] A retired woman from southern Ontario, she was on a flying tour of northern Canada. She referred to this vacation as "the trip of a lifetime" and was thrilled to tell me about her adventures. When asked about her feelings about the trip so far, Janice paused, took a deep breath, and with tears in her eyes and hand on her chest, said:

> Let's just say I'm proud to be Canadian. The land. It is so big and vast and amazing. I've always felt like the North was like the soul of the Canadian land, but now I see it more as the heartbeat.

Scanning the view from where Janice and I are sitting, I see several meticulously restored sites that are part of the Dawson Historical Complex National Historic Site of Canada. Walking the streets of Dawson City, I come across the others that make up this complex of seventeen sites of officialized national heritage (see the city's website for an excellent visual overview of these sites: http://www.dawsoncity.ca/klondikeattractions/parkscanadahistoricalsites/). Parks Canada calls for a cooperative relationship with all parties in an effort to manage Dawson's overall visual presentation and maintain "the spirit of the place."[17] In the spaces in between the official sites of national heritage, municipal bylaws regulate the aesthetic presentation of Dawson City's streetscape by ensuring that all buildings conform to Parks

Canada-mandated architecture designed to "evoke the image of a gold rush town."[18] Official heritage commemorations in the form of restored buildings make sense only when they are utterances expressed within a coherent heritagescape. There are coherent national-cultural and visual narratives *taking place* around us, designed to evoke in us a sense of national-cultural affection. Janice's words flow out of, and back into, this broader national-cultural story as she narrates herself into the text as a proud Canadian national-cultural subject.

National-cultural imaginaries are guided, in part, by dominant fictive narratives.[19] These are the national stories told about the nation, its past, present, and future. As mentioned earlier, the story of Dawson City as it is dominantly told stands in as emblematic of the story of Canadian nation building. Official historical commemorations work as technologies of place making to spatialize and ground these narratives.[20] They serve to transfer and translate dominant national-cultural narratives into grounded physical places and in so doing inscribe space with social and historical meanings and power relations.

Geographer Brian Osborne terms this process of national-cultural identification as the cultivation of an "a-where-ness."[21] Official commemorative efforts work to compress time and space to produce national places.[22] This is accomplished through nurturing national-cultural metanarratives and grounding these narratives materially in fixed sites of commemoration.[23] The material landscape of the nation becomes "loaded with symbolic sites, dates, and events that provide social continuity, contributing to the collective memory, and established spatial and temporal reference points for society."[24] The goal of this process is to ground national-cultural imaginaries whereby

> national identities are coordinated, often largely defined, by legends and landscapes, by stories and golden ages, enduring traditions, heroic deeds and dramatic destinies located in ancient or promised homelands with hallowed sites and scenery. The symbolic activation of time and space ... gives shape to the imagined community [the Nation].[25]

In Dawson City, this official spatialization commemorates the Klondike Gold Rush for its contribution to Canadian history. The story of Dawson is produced, told, and spatialized as the story of Canada. Parks Canada, the federal governmental agency responsible for the maintenance of these sites, outlines three main features of the Klondike Gold Rush's significance that need to be highlighted through commemorative efforts:

- Dawson as a reflection of the character of the Klondike Gold Rush and its impacts.
- The impact of the Gold Rush on the development of the Yukon Territory.
- The impact of Dawson and the Gold Rush on the collective imagination of Canadians.[26]

The goal of heritage production in Dawson City, as outlined by Parks Canada above, is to foster a sense of national-cultural "a-where-ness" that ties the local to the regional and the regional to the national. Dawson City matters in Canadian national-cultural imaginaries because of the ways that it, and the Klondike Gold Rush, represents in national-cultural metanarratives: wilderness discovered, riches found, hardships endured, and civilization established. Of this process, former Parks Canada historian David Neufeld says:

> The commemoration of the Gold Rush has little to do with the Gold Rush itself and everything to do with development of natural resources across northern Canada in the fifties and sixties. And so it is a commemoration, as all commemorations are, set in its time. And so the romance of the Gold Rush is replicated in order to excite the public, or buy them into the whole sense of the North and the development into the North and the exploitation of natural resources ... It has little to do with the gold rush itself, except as a vehicle for this other story.[27]

Dawson City fits well into this guiding narrative structure. The Klondike Gold Rush is easily retold as a familiar tale of conquest, fortitude, national expansion, and romance. It hooks into established and well-told national-cultural imaginings of the North and nation building. It fits into and flows out of the broader ways in which history has dominantly been written, and heritagescapes produced, in Canada. Visitors to Dawson City can tie a pre-existing sense of national selfhood onto the story of the Klondike Gold Rush. It makes sense because it is an articulation of an already well-told national tale – a tale told on, in, and through the imaginative terrain of a northern wilderness frontierscape that sparkles with the image of gold flecks lining creek beds, free for the taking.

Politics in Time and Space

State-sponsored and officialized heritage sites are cultural artifacts, not of what they assume to commemorate, but of the politics and processes by which such commemorations come to be commemorated.[28] In 1946, not long after the Soviets looked to Canada for guidance with their colonial project,[29] Canadian prime minister L.B. Pearson nodded back, declaring that "Canada, like Russia, is looking to the North as a land of the future."[30] Canada's next prime minister, John Diefenbaker, picked up this "Northern Vision," and it continued to gain political and emotional momentum into the 1950s. Diefenbaker centred his successful campaign for prime minister in 1958 on this "great Northern Vision," which led to the largest majority government in Canadian history at the time. In 1959, Diefenbaker travelled to Dawson City and "himself first raised the possibility of developing Dawson City as an historic tourist attraction ... he pointed out the attraction of the heritage gold rush for Yukon visitors."[31] The result of this visit was the initiation of the process to officialize national heritage in Dawson City. Of this process David Neufeld says:

> It was very much coming straight from the Prime Minister's Office. Somewhat unusual, but this was Diefenbaker's time. This was the Northern Vision and this is what they were going to do ... the reason John Diefenbaker pushed this and the federal government continues to support it is that Dawson was dying. Gold mines were shutting down. And if you have a Northern Vision, you need to have ... well, having a town die in the north doesn't work ... if the north is the future, we are going to resurrect it through this process of commemoration.[32]

Three points are significant here. The first is that Diefenbaker saw Dawson City as a symbol of his broader northern national aspirations. From the outset, the production of heritage in Dawson City was tied to, and deployed, as a technology in a larger national political project. Next, Diefenbaker identified the Klondike Gold Rush as the "significant" historical event in Dawson City. It was this event that would serve as the central node upon which "historical significance" would be defined (back-filled) in Dawson.[33] And third, from its very inception in Dawson City, heritage was linked to tourism. These three points form the platform upon which heritage as a state-sponsored project came to be produced in Dawson. The Klondike Gold Rush would act as the axis around which heritage would be produced.

Heritage would be produced as a conduit for the articulation of national northern achievement, expansion, and promise. And heritage in Dawson City would be produced specifically as a tourist attraction. The making of Dawson City as a place of national historical significance and a tourist destination occurred (and occurs) at the confluence of these three goals.

Capitalizing on the lure and lore of these imaginative nuggets, official heritage in Dawson City was born out of a particular political and economic present and was assigned a specific job. Official heritage production was deployed as a technology of placemaking designed to pull the Klondike Gold Rush out of the historical past and infuse it with meaning in the present. This was designed to forge connections between carefully selected "then" and a constantly shifting now. National imaginings of the North as a resource-rich hinterland space are mobilized, and the gold rush is transformed into an emblematic expression of this imaginary. The past, in this case the Klondike Gold Rush, is back-filled with meaning to meet the political and economic needs of the present. Well-known scholar of cultural heritage management Gregory Ashworth suggests:

> Heritage is by definition an inheritance from the past destined for the future. It is a product of the present purposefully developed in response to current demands for it, and shaped by that market. It follows that the past is a quarry of possibilities from which selection occurs not only, or even principally by chance survival but by deliberate choice.[34]

Ashworth is pointing to the selective, subjective, and political processes that underscore the production of heritage. He is also highlighting the importance of the present in recuperating renditions of the past to be called "heritage." On this note, performance studies scholar Barbara Kirshenblatt-Gimblett writes:

> Heritage is not lost and found, stolen and reclaimed. Despite a discourse of conservation, preservation, restoration, reclamation, recovery, re-creation, recuperation, revitalization, and regeneration, heritage produces something new in the present that has recourse in the past.[35]

Heritage can thus be understood as a discursive process of invention, a technology, underscored by contemporary political desires, economic interests, and established power relations. In Dawson City, a prime minister with ambitious visions for Canada's northern territories provided the political context out of which heritage-as-national-vision-as-tourism-industry emerged. His desire merged with increased popular (and positive) imaginings of this imagescape called the North. On the ground in Dawson City, these forces converged and a national-cultural heritagescape as tourist attraction emerges.

The process of seeking and establishing official National Historic Site designation is a process of symbolification, or encoding national-cultural guiding fictions and allowing them to congeal in place. The Klondike Gold Rush is a historical moment that is produced as heritage, and this heritage is articulated by way of the recreation of Dawson's built streetscape. Officialized national history is spatialized and Dawson City emerges, happens, as a *place* of national-cultural significance and interest. The gold rush on its own is not the point of interest. It gains momentum as historically significant only by way of linking it to a larger national historical narrative *and* by drawing people north to tour its significance. Geographer Michael Pretes suggests that "the monumentification of a site or thing transforms intangible fictions into somatological, visual, and consumerable forms. They are material articulations of nationalist ideals that also serve as an object of pilgrimage."[36] Getting people to travel to Dawson City was a key part of the project from the start. Canada's northern vision needed to resurrect Dawson City as a thriving community to hold its credibility.

Discourses and practices of tourism join forces with officialized national commemorative efforts to transform the streetscapes of Dawson City into a readable, understandable, and tourable northern environment. This North, the one imagined and made visible in Dawson City, is the vision of the frontier – a resource-rich space on the edge waiting to be discovered, conquered, and claimed by those strong and brave enough to endure it. It's the Wild West northwarded. Tourism was part of the equation in the production of heritage in Dawson City from the beginning. This means that national-cultural heritage was, and is, produced in Dawson City in the image of both touristic and nationalist desires. The aesthetic presentation of the past in Dawson City is the product of a discursively dense present that is shaped by, and shapes, broader notions of both national-cultural and touristic imaginations of the North. Images of Dawson City as a knowable, readable national-cultural heritagescape – how people imagined a northern gold-rush-frontier town – are one thread in this knot. What those images of the North speak to in relation to drawing people into a national-cultural imaginary is another. The discourses and practices of tourism tighten around both of these threads, and Dawson City comes into view as a specifically northern place in carefully selected terms.

Tourism studies scholar Dean MacCannell argues that tourism should be understood as a restless quest for authenticity.[37] I would argue that tourism, rather, produces its own complex of meaning that defines authenticity. One visitor to Dawson who I spoke with declared:

> Doesn't it feel real? The dirt streets and old buildings really give an authentic feel. It's like the Wild West in real life.[38]

For this visitor, the presentation of Dawson City followed familiar, recognizable tropes, as a heritage site *and* tourist destination. This visitor felt the realness of Dawson based on its aesthetic presentation and her imaginings she brought with her of what that realness should look and feel like. Dawson City was recognizable as "the Wild West in real life" because of the ways that the production of heritage matched her ideas of what the Wild West looked and felt like.

In another conversation, Barb, a woman from British Columbia, declared, "It's just so authentic! I love it!"[39] We met in the campground in downtown Dawson City late one afternoon. The comment came directly after I asked about her "Dawson experience" and Barb replied:

> I am having so much fun here. Last night I went to Gerties.[40] I'm going again tonight. I just got my tarot cards read. And my husband is out getting a massage right now. I'll have one tomorrow. I can't believe that we travelled so far through so much nothing to find this little gold nugget [laughing]. There is everything to do here. It's fabulous.

These comments contain what first appear as a host of contradictions. Is having tarot cards read and getting a massage an authentic northern gold-rush town experience? Gambling, dance-hall girls, spa treatments, and gold-nugget souvenir shopping all form Barb's Dawson experience as an "authentic" touristic adventure.

Parks Canada, the City of Dawson, and other agents of heritage production in Dawson hold as their primary concern historical authenticity. This is an invented authenticity, the definition of which flows out of fictive narratives and collides with a set of touristic expectations about what makes up an authentic tourist destination. For Barb, Dawson City is authentic to its own construction of what an authentic northern gold-rush town-as-tourist-attraction should look and feel like. Barb's Dawson experience had also included panning for gold, eating "a great meal" at Klondike Kate's, watching the dance-hall girls at Diamond Tooth Gerties, shopping for gold jewellery and postcards, and taking a ride around town on a horse-drawn carriage. Complexes of discourses and practices converge to produce the conditions by which notions of authenticity are legitimated in Dawson City.

I bring up the question of authenticity here not to debate the concept but rather to confront how power operates through it. David Neufeld suggests that as it operates in the making and maintenance of national heritage in Dawson City, authenticity is conflated with authority.[41] This means that assertions of authenticity in how Dawson City is presented are exercises of power. In Dawson City, heritage *takes place* in the built form of the town, and how this heritage is produced and articulated is regulated by those charged with determining what is authentic and what is not. By deploying official heritage production as a technology of place making, power relations are literally built into Dawson City's streetscape-as-heritagescape.

As statements from visitors to Dawson City cited throughout this chapter suggest, Parks Canada and the City of Dawson are largely successful at creating a place that feels authentic to what people expect a National Historic Site, tourist destination, and northern environment to look and feel like. Dawson City presents itself, at first glance, as a charming site of national-cultural northern historical significance and pride. The trouble is that history is messy, uncomfortable, and complicated. People on holiday are not always interested in confronting the ghosts haunting the margins of their senses of national selfhood. Tying the production of heritage in Dawson City to a broader national narrative of nation building requires softening the blows of those things done in the name of that project. This is the magic of heritage. It is produced, edited, celebrated, and revised as an ongoing, emergent product of the present. In Dawson City, heritage is produced as a romantic backwards glancing at a national-cultural point of origin. It encapsulates a long-distance love and turns this love into an industry. The production of official heritage is fuelled by, and in turn fuels, national-cultural imaginings. Cultural logics converge and heritage is produced as simultaneously an imaginative project and economic industry. In Dawson City, these logics come together to allow Dawson City to *take place* in the image of narrative intension, nationalist desire, touristic consumption, and colonial power relations – which then raises the question, if travel north is for so many an enactment of national-cultural subjectivities, is it not also a moment where one must confront those things done in their names as a national-cultural subject?

An Other(ed) View from the Banks of the Klondike River

Remember Janice, the woman from Ontario who expressed an emotional attachment to a sense of Canadian-ness as a result of her travels north? What I did not mention earlier is that Janice and I met on the porch of the Dänojà Zho Cultural Centre, situated on the banks of the Klondike River on Front Street, in downtown Dawson City.[42] It was here that she spoke so passionately about her affections for her national-cultural home. It was here that Janice spoke the words:

> Let's just say I'm proud to be Canadian. The land. It is so big and vast and amazing. I've always felt like the North was like the soul of the Canadian land, but now I see it more as the heartbeat.

She continued:

> I travel all the time. This year alone I'll go to India, China, and Australia, as well as this trip. But after this experience I feel more Canadian, or feel more what it feels to be Canadian, if that makes sense?

As I looked at Janice, in this moment, over her shoulder I could see the haunting image of a child's face on a poster hanging in the window of the cultural centre. It was a poster for the exhibit inside – "Where Are the Children? Healing the Legacy of the Residential Schools." The walls of the exhibit hall were filled with images of indigenous children at residential schools all over Canada (the project's website, http://www.wherearethe children.ca/en/exhibit/, presents the photographs and the stories around them). Poster panels of first-hand accounts of local Tr'ondëk Hwëch'in First Nations citizens' recollections of their experiences at some of these schools flanked the images. Of the layout of the exhibit, manager Glenda Bolt said:

> Some of the residential school survivors that have spoken to me expressed the feeling of being trapped. They were alone with no way out, confused by all the new and strange things and consumed by feelings of being overwhelmed and scared to the core. I wanted to find a way to incorporate those intense emotions in the exhibition to make sure that people realized that we

> are speaking about children ... powerless, confused little kids that were far from home and help. I purchased free-standing blackboards and created a maze of sorts to hang the framed black and white photographs on ... all hung with industrial-looking wire and hooks fashioned from metal brackets. I wanted it to have the feel of an old-style classroom but designed in such a way that almost by accident you realize that you are stuck in a corner, surrounded by the images and words, with the sound of children on the sound system overhead and not able to see a clear path to the exit. Trapped.[43]

Free-standing blackboards served as display walls arranged in such a way as to trap people moving through the exhibit. Photographs were displayed on the walls and blackboards. Panels of text were used to contextualize, and personalize, the photographs. The passages of text were taken from oral histories told by Tr'ondëk Hwëch'in survivors of residential schools. They talked about life before school, memories of the day they were taken off to school, their first experiences there, their time away at school, and the difficulties of coming back to their community after that time spent away. These passages were eloquent, powerful, chilling, and brave. The photographs were beautiful black and white images of children and schools from across Canada. Janice and I happened to walk out of the exhibit at the same time.

As Janice and I spoke on the porch, the image of the child's face over her shoulder haunted her words. Through her first-hand travel north, Janice felt a growing sense of attachment to her national-cultural home. But whether she knew it or not, this love could survive only by her turning her back on the children whose images hung inside.

National-cultural imaginaries are haunted by the pasts and violences they work to forget. Nations are narrated through a complex process of remembering and forgetting. Nations celebrate and remember antiquity and forget the violences out of which they emerged.[44] As English literature and post-colonial scholar Homi Bhabha so eloquently suggests, "It is this forgetting that constitutes the *beginning* of the nation's narrative."[45]

So, as official state-sponsored national heritage works to *place* national-cultural memories on the ground in Dawson City, it also actively works to forget those historical bits that complicate the smooth, heroic surfaces of the story. In Yukon, these dominant national-cultural narratives revolve around celebrations of colonial encounters and victories. Airplane, roads, railways, and narratives are deployed as technologies in this effort. The North needs to be imaginatively made as an unpeopled wilderness edge, full of riches, waiting to be discovered – a frontier – in order to be claimed as such. That the North continues to be activated in dominant Canadian national-cultural imaginaries in these terms suggests that the work of conquest continues. That so many travel north to Dawson City to celebrate these colonial accomplishments suggests that we are all active agents in this ongoing colonial and colonizing project.

The child in the poster over Janice's shoulder reminds us that the official rememberings, and their accompanying forgettings, *taking place* around us in Dawson City are haunted. Sociologist Avery Gordon suggests that hauntings are those moments "when the over-and-done-with comes alive, when what's been in your blind spot comes into view."[46] Gordon goes on to state that hauntings, like the face of this child over Janice's shoulder,

> [are] one way in which abusive systems of power make themselves known and their impacts felt in everyday life, especially when they are supposedly over and done with (slavery, for instance) or when their oppressive nature is denied (as in free labor or national security).[47]

This child's face haunts Janice's words because it brings back to life "the over-and-done-with" and "demands its due."[48] We cannot talk about the ways in which the North *takes place* in Dawson City without talking about the ongoing colonial legacies and relationships that that process celebrates, allows, energizes, and reconstitutes. These power relations *take place* too, and the child in the poster over Janice's shoulder reminds us all that she and all her relations demand their due.

As mentioned in the opening of this chapter, if the story of Dawson City is meant to stand in for the story of Canada, then it is a story of colonization and the brutalities of those things done in the name of Canadian nation building. The founding colonial principal that guided nation building in Canada was the forced assimilation of indigenous occupants of the territories now known as Canada into the newly established body politic. Several strategies converged in this effort, including outlawing many indigenous cultural practices, forced relocations of indigenous communities, the allocation of Indian reserves, and the creation of a registration system determining who was and was not an Indian – and then the creation of a series of ways to systematically reduce the numbers of registered Indians.[49] One of the most aggressive of these strategies was the residential school system, which was designed to speed the assimilation process by removing indigenous children from their families and communities and placing them in boarding schools. Attendance at residential schools was first legislated as mandatory. Later this legislation was strengthened, making it illegal not to send your children, thereby criminalizing acts of non-compliance with increased formal legal sanctions and punishments; thus, most indigenous children in Canada between the 1890s and the late 1960s spent the majority of their childhoods in these residential schools. The goal of these schools was not education but forced assimilation. The result was that many children over several generations suffered brutal sexual, physical, and emotional abuses.[50]

The effects of the residential school system on indigenous peoples in Canada cannot be understated. Generations of children were removed from their homes, and communities all over Canada devastated. It is reported that because of overcrowding, undernourishment, and poorly heated and maintained buildings, as many as 50 percent of all children who entered the residential schools did not survive to return home. Many of those that did survive returned home to communities devastated by the compounding effects of colonialism, where alcoholism, unemployment, and despair from the loss of so many had taken hold. The residential schools became a catchment basin for children whose communities were devastated by colonialism, in part because of the effects of these very schools. Referred to as a holocaust for indigenous peoples in Canada, the suffering and destruction to both individuals and communities that resulted from the residential schools is an integral part of the story of Canada.[51]

This part of the story of Canadian nation building does not fit neatly with meticulously restored gold-rush-era buildings and gold-nugget souvenir shops. Technologies of placemaking are also technologies of power. The repetitive insistence on the importance of the Klondike Gold Rush in the production of Dawson City continuously attempts to silence out any other versions of the story. The Dänojà Zho Cultural Centre interrupts this silence. The child's face in the window haunts it.

Perched on top of the dike that runs along Front Street in Dawson City, a postmodern structure stands out – it looks out of place. Set among the false fronts, wooden boardwalks, gold-nugget souvenir shops, and Parks Canada interpretive guides walking the streets in petticoats and long skirts, this impressive building interrupts the narrative flow of this heritagescape (see http://trondekheritage.com/danoja-zho/tour-the-centre/ for photos of the building). The Dänojà Zho Cultural Centre is a post-colonial assertion on a colonial national-cultural landscape.[52] And it's unsettling. One of Dawson City's central appeals as a tourist destination is that it is produced and maintained in recognizable, comfortable terms. National-cultural (colonial) nostalgias are activated, energized, and *placed*. The Dänojà Zho Cultural Centre unsettles the very terrain, literally and figuratively, upon which these national-cultural imaginaries operate, *take place*, in Dawson City. In so doing, Dänojà Zho dislodges Dawson City's, and Canada's, colonizing pasts and present from nostalgia's comforting embrace.

Many of the visitors I spoke to over the course of three summer seasons in Dawson City found this interruption troubling. Many were unsettled by the structure's difference set against what they felt was the very "authentic" historical recreation of Dawson City's streetscape. As I stood on the bank of the river one afternoon, talking with a visitor from Alberta, our conversation moved toward the "feel" of Dawson. This traveller recounted some of his thoughts about what he expected Dawson City to look like and declared that, for the most part,

> it really is what I expected. It's like being in a postcard, or an episode of *Deadwood* ... except for that building [pointing to the Dänojà Zho Cultural Centre]. It just looks out of place. I don't get it.[53]

This visitor may not "get it," but he certainly pointed out a key feature of the structure. It is "out of place" by design. The Dänojà Zho Cultural Centre merges contemporary design with traditional Tr'ondëk Hwëch'in architectural features. Of the architectural design of Dänojà Zho manager Glenda Bolt states:

> Getting the architecture right was so important. The committee had to find a way to reflect Tr'ondëk Hwëch'in culture and values, and it became obvious rather quickly that, whatever was designed, it would not adhere to the gold-rush streetscape theme ... "Dawson style." The centre's architecture does reflect 1898, but just not the false-front style of the stampeders. It reflects the shelter and camps of that time period at Tr'ochëk fish camp. It's been a hard one for some Dawsonites and visitors to deal with and figure out; people are very gold-rush-theme-park-oriented here. The First Nation has constructed plenty of buildings that adhere to the planning board bylaws, including their homes and government office buildings ... just not the cultural centre. People want to get along and work together, that's a true value of this community, is people getting along. But with this particular building, the line was drawn when it came to the exterior.[54]

The structure's out-of-placeness forces a gap in the production of place in Dawson City and the placement (or displacement, as the case may be) of indigenous peoples in the story of Dawson City. By opening a gap on the national-cultural and touristic landscapes of Dawson City, Dänojà Zho re-places Tr'ondëk Hwëch'in people in place. The result is the emergence of a space where "an-Other" story of Dawson can *take place* and where the North as a terrain of national-cultural ambitions and desires needs to be confronted not just for what it is but for what it does.

Imaginaries Matter

How the North gets imagined matters. National-cultural imaginaries do more than shape the ways in which spaces are imaginable. Northern environments, places, happen in their image. Discursive technologies deployed in the creation of northern environments as resource-rich hinterlands renew energy for their exploitation. The production and celebration of Dawson City as a site of national historical significance because of its resource-rich history serves the interests of contemporary economic desires for resource extraction activities in other northern places. The narrative of the North as a terrain upon which riches can be discovered is energized. The impacts of this can be seen throughout the circumpolar north as other northern environments come into view as ripe for exploration and exploitation.

As part of this same discursive process, colonial relations too remain and are reaffirmed. The Klondike Gold Rush emerged into national historical significance fifty years after it ended. Set in a particular politics of the present (in the 1950s), John Diefenbaker scanned Canada's historical landscape for moments that could be easily reduced to symbols of national-cultural aspirations for northern development. The Klondike Gold Rush came into view as one such moment, and its significance to national history was back-filled through the production of official state-sponsored heritage. The Klondike Gold Rush emerges as a celebrated symbol of national-cultural vision, ambitions, and character. Dawson City is produced, *takes place,* as a site of national historical significance and a tourist destination in these colonial and colonizing terms.

Power relations get built into landscapes, streetscapes, and places and then disappear out of sight, allowing them to go unquestioned. The historically specific colonial power relations between indigenous and non-indigenous peoples in the region underscore the entire tale of tourism and national-cultural heritage production in Dawson City. Celebrating the

Klondike Gold Rush is celebrating Canada's great colonial achievements and everything done in their wake.

This discussion ends deliberately at the Dänojà Zho Cultural Centre because of the potentials opened up by the interruption that it forces. In a decidedly non-confrontational way, Tr'ondëk Hwëch'in citizens interrupt national-cultural pride and in so doing force a critical look not just at how ideas of the North *take place* in Dawson City, and what power relations are maintained and reconstituted in the process, but also at what new configurations of culture and power might be possible. As the articulation of the totality of relations in any one moment, northern environments are always emerging, changing. By facing the child in the poster over Janice's shoulder, really looking at her, welcoming her in the present, honouring her, a space of potential opens up for a rethinking of how the North, Yukon, and Dawson City can *take place*. From this emerging present there is hope to "transform a shadow of a life into an undiminished life whose shadow touches softly in the spirit of a peaceful reconciliation."[55] To the child in the poster, and the courage and generosity of the Tr'ondëk Hwëch'in, I offer a heartfelt *Mähsi Cho* (big thanks). It is my hope that through the spirit of peaceful reconciliation we can, from all sides of this post-colonial project, transform the shadow of a life into an undiminished one and collectively start to reimagine how this post-colonial North might *take place*.

NOTES

1 Akhil Gupta and James Ferguson, "Culture, Power, Place: Ethnography at the End of an Era," in *Culture, Power, Place: Explorations in Critical Anthropology*, ed. Akhil Gupta and James Ferguson (Durham and London: Duke University Press, 1997), 1-29.

2 Bruce Braun, "Place Becoming Otherwise," *BC Studies* 131 (2001): 19.

3 Yukoninfo.com, http://www.yukoninfo.com/dawson/.

4 Kenneth Coates and William R. Morrison, *Land of the Midnight Sun: A History of the Yukon* (Montreal and Kingston: McGill-Queen's University Press, 2005); Melody Webb, *Yukon: The Last Frontier* (Vancouver: UBC Press, 1993).

5 There is some confusion over the name of the man from Tagish who has become known as both Tagish and Dawson Charlie. Although some accounts refer to him as Tagish Charlie (Webb), he was inducted into the Canadian Mining Hall of Fame in 1999 as Dawson Charlie (http://mininghalloffame.ca/) and is often referred to by this name in historical accounts of the Klondike Gold Rush (Coates and Morrison, *Land of the Midnight Sun*), 79; Webb, *Yukon*, 123.

6 Kathryn Morse, *The Nature of Gold: An Environmental History of the Klondike Gold Rush* (Seattle and London: University of Washington Press, 2003).

7 Ibid.
8 Webb, *Yukon*, 129.
9 Most journeyed to the Klondike by way of the Alaskan coast, but some did travel an inland route from Edmonton, Alberta. Coates and Morrison, Morse, and Webb all refer to this alternate route in more detail.
10 Morse, *Nature of Gold*.
11 Coates and Morrison, *Land of the Midnight Sun*, 105-11. Dawson City remained the capital of Yukon Territory until 1953, when the designation was moved to Whitehorse.
12 Tr'ondëk Hwëch'in Heritage Sites, http://trondekheritage.com; ibid., 110-12.
13 See Morse, *Nature of Gold*, for a detailed environmental history of the Klondike Gold Rush.
14 Bruce Braun, *The Intemperate Rainforest: Nature, Culture, and Power on Canada's West Coast* (Minneapolis: University of Minnesota Press, 2002), 19.
15 Rob Shields, *Places on the Margin* (London: Routledge, 1991); Sherill E. Grace, *Canada and the Idea of the North* (Montreal and Kingston: McGill-Queen's University Press, 2001).
16 All names have been changed and pseudonyms used to respect and protect the anonymity and confidentiality of those generous enough to share time with me throughout this research.
17 Parks Canada, "Dawson Historical Complex National Historic Site Commemorative Integrity Statement," *Design Guidelines for Historic Dawson*. Prepared by Environmental Services Division, Engineering and Architecture Branch, Finance and Professional Services, Ottawa, and Restoration Services Division, Parks Canada, Prairie Region, Winnipeg, n.d., 6.
18 Ibid., 4.
19 M. Pretes, "Tourism and Nationalism," *Annals of Tourism Research* 30 (2003): 125-42.
20 Joanna Breidenbach and Pál Nyíri, "'Our Common Heritage': New Tourist Nations, Post-'Socialist' Pedagogy, and the Globalization of Nature," *Current Anthropology* 48 (2007): 322-30.
21 B.S. Osborne, "Landscape, Memory, Monuments, and Commemoration: Putting Identity in Its Place," *Canadian Ethnic Studies* 33 (2001): 39-77.
22 Rhys Jones and C. Fowler, "Placing and Scaling the Nation," *Environment and Planning D* 25 (2007): 332-54.
23 Pretes, "Tourism and Nationalism."
24 Osborne, "Landscape, Memory, Monuments, and Commemoration," 39.
25 Stephen Daniels, *Fields of Vision: Landscape Imagery and National Identity in England and the United States* (Princeton, NJ: Princeton University Press, 1993), 5.
26 Parks Canada, 5.
27 Personal communication, June 2008, Whitehorse, Yukon. Cited with permission.
28 C.J. Taylor, *Negotiating the Past: The Making of Canada's National Historic Parks and Sites* (Montreal and Kingston: McGill-Queen's University Press, 1990), x.
29 Julia Lajus, this volume.

30 Lester B. Pearson, "Canada Looks 'Down North,'" *Foreign Affairs* 24 (1946): 638-47.
31 Taylor, *Negotiating the Past*, 171.
32 Personal communication, June 2008, Whitehorse, Yukon. Cited with permission.
33 David Neufeld, "Public Memory and Public Holidays: Discovery Day and the Establishment of a Klondike Society," *Going Public: Public History Review* 8 (2000): 74-86.
34 G. Ashworth, *Let's Sell Our Heritage to Tourists?* (London: Council for Canadian Studies, 1994), 1.
35 Barbara Kirshenblatt-Gimblett, *Destination Culture: Tourism, Museums, and Heritage* (Berkeley: University of California Press, 1998), 149.
36 Pretes, "Tourism and Nationalism," 133.
37 Dean MacCannell, *The Tourist: A New Theory of the Leisure Class*, 2nd ed. (Berkeley and Los Angeles: University of California Press, 1999).
38 Interview with author, Dawson City, Yukon, 20 July 2008.
39 Interview with author, Dawson City, Yukon, 18 July 2007.
40 Diamond Tooth Gerties Gambling Hall is a casino run by the Klondike Visitors Association.
41 Personal communication, Whitehorse, Yukon, January 2009. Cited with permission.
42 The Dänojà Zho Cultural Centre is a centre operated by the Tr'ondëk Hwëch'in First Nation in Dawson City. It is located on Front Street, on the banks of the Klondike River.
43 Personal communication, Dawson City, Yukon, July 2008. Cited with permission.
44 Michael Billig, "Banal Nationalism," in *Nations and Nationalism: A Reader*, ed. Philip Spencer and Howard Wollman (New Brunswick, NJ: Rutgers University Press, 2005), 184-96.
45 Homi Bhabha, "DissemiNation: Time, Narrative, and the Margins of the Modern Nation," in *Nation and Narration*, ed. Homi Bhabha (New York: Routledge, 1990), 291-322, quotation at 310.
46 Avery F. Gordon, *Ghostly Matters: Haunting and the Sociological Imagination*, 2nd ed. (Minneapolis and London: University of Minnesota Press, 2008), xvi.
47 Ibid.
48 Ibid.
49 Olive Patricia Dickason and William Newbigging, *A Concise History of Canada's First Nations*, 2nd ed. (Oxford and New York: Oxford University Press, 2010).
50 Suzanne Fournier and Ernie Cray, "'Killing the Indian in the Child': Four Centuries of Church-Run Schools," in *Racism, Colonialism, and Indigeneity in Canada*, ed. Martin J. Cannon and Lina Sunseri (Oxford and New York: University of Oxford Press, 2011), 173-76.
51 Bonita Laurence, *"Real" Indians and Others: Mixed-Blood Urban Native Peoples and Indigenous Nationhood* (Vancouver: UBC Press, 2004).
52 Post-colonialism, as I am using the term here, is understood as a critical project that traces "the cultural, economic, and political conditions that exist in the aftermath of colonialism and names a desire to engage them in a critical fashion": Braun, *Intemperate Rainforest*, 21.

53 Milch, *Deadwood* (HBO, 2004); interview with author, Dawson City, Yukon, 20 July 2007.

54 Personal communication, Dawson City, Yukon, July 2008. Cited with permission.

55 Gordon, *Ghostly Matters*, 208.

Iceland and the North
An Idea of Belonging and Being Apart

UNNUR BIRNA KARLSDÓTTIR

Iceland is located on a hot spot in the North Atlantic Ocean, on the geologic rifts between the Eurasian plate and North American plate, hence its eruptions and earthquakes. The best-known natural characteristics of Iceland are the volcanic activity and the geothermal sites with hot springs and geysers. Mountainous peninsulas, fjords, and bays shape the island's outlines. The habitable area is the lowland around the coast and into the valleys, with its small towns and farmland. In the centre lies the highland. It is a desert for the most part, and mountains and glaciers rise up to the sky. There are also vegetated zones, a few wetland oases, glaciers, large areas covered with lava, and glacial rivers that stream down to the sea.

The Icelandic nation is still mostly homogenous, with ancestral roots in Norway, the British Isles, and Ireland, as the first settlers came from there. The number of inhabitants in Iceland today is just above 300,000.

The idea of the North is perceived in Iceland as a definition of the country's place in relation to the rest of the world, in both a geographical sense and a cultural sense. Thus, the North is a geographic space and a cultural construction. Icelanders generally use the term "North" to define a region or parts of a region, depending on the context. It can mean one, some, or all of the following: the Arctic, Greenland, Iceland, Denmark, Norway, Sweden, Finland, and the North Atlantic Ocean. It also refers to the climate in the northern hemisphere and to the inhabitants of certain countries, stretching from Greenland and Iceland in the west to Finland in the east, and everything in between.

Ideas of nature and the North are also so tightly intertwined in Icelandic contemporary history that these terms often merge in public discourses. So the North means nature and nature means the North. This is partly because of the image of the Nordic countries as places with the largest wilderness areas in Europe, compared with the more cultivated and densely populated European continent.[1] In addition to this, Iceland is seen as a world apart in the North, thanks to it being an island in the North Atlantic Ocean, quite far from any other country. Being on an island in the middle of the North Atlantic Ocean has always marked the lives and perceptions of Icelanders – modern technology and the values of industrialized consumer society in the twentieth and twenty-first centuries have not changed that. Living on an island in the northern hemisphere has a clear effect on Icelandic identity, including Icelandic mentality, society, politics, culture, economy, resource exploitation, and environmental issues. Thus, it is not only geographical and geological factors that make Iceland a world apart but also an ideology, as I portray here. My focus is on ideas about the relationship between people and nature in Iceland in the period 1900-2008, and I use the debate on hydropower projects in recent years to elucidate that story.

The Cultural "North" – An Idea of Belonging

The idea of a nation and its relationship with the nation-state, and its right to be such a state, became hegemonic in the Icelandic political discourse in the second half of the nineteenth century and firmly rooted among the general public in Iceland in the first half of the twentieth century. But Icelanders did not only strive from about 1900 to acquire their own independent nation-state – to secede from Denmark, and indeed, succeeding in 1944 – they also wanted to be regarded as a part of the community of Nordic nations.[2] Thus, Iceland wanted to stand apart, as a sovereign nation, yet belong to the family of Nordic nations.

This vision of the importance to be on equal terms with the other Nordic nations, economically, politically, and culturally, and to be seen as a part of their world, contained the idea of belonging to a place stretching further than the geographical borders of Iceland and the national borders of the Scandinavian countries. In this sense, the term "North" describes a culturally constructed space where "the North" stands for a group of nations seen as a world apart from other nations, in both European and international contexts. Icelanders have, therefore, since the latter half of the twentieth century, laid an importance on participating in Nordic cooperation at the cultural, social, and political level. In the minds of Icelanders, this cooperation has had a certain kind of priority over relations and cooperation with other nations. The reason this is so is that Icelanders today connect the idea of the North as the Nordic nations with the idea of historical and cultural ties, since Iceland was settled by Scandinavians, and the country was ruled by Norwegian kings from 1262 and then Denmark from the fourteenth until early twentieth century.[3]

Language is a cultural link in Scandinavia. The Icelandic language is closely related to Norwegian, Danish, and Swedish, but most Icelanders use English to communicate with their Nordic neighbours. This is because of the strong influence of English and the Anglo-Saxon culture in Iceland since the Second World War and also the strong position of English for the last few decades as the international language. English is the second language Icelandic teenagers learn in school today. English has superseded Danish, which was learned as the second language until late in the twentieth century, given Iceland's history with Denmark.[4] Still, Icelanders generally prefer to identify themselves with the other Nordic countries when it comes to basic social values. Although Iceland has had conservative right-wing governments for different periods in the recent decades, there has nevertheless been a tendency among Icelanders in general to see the Scandinavian model of the social-democratic state as more appealing and closer to the Icelandic cultural and political character than the English or American capitalistic one. This builds on Iceland's left-wing and social-democratic political history, and the history of the construction of the Icelandic welfare state, both very much influenced by the development of a social-democratic state in Scandinavia.[5] This was, for example, demonstrated after the collapse of the Icelandic banks in 2008 in the rising public demands that Iceland not continue following the neo-liberal path that had led to such a grand-scale financial disaster but instead turn toward social-democratic values in restoring the Icelandic economy and society.[6]

So, "the North" as a term to define the Icelandic nation's place in the community of nations is an important thing for Icelanders. It gives them a certain kind of identity and a definition – that they belong to the community of Nordic nations and are therefore a part of the world that endorses certain standards when it comes to human rights, democracy, and human welfare. In this sense, the North is an ideological place, and a term that defines something of a cultural, political, and social worth.

Nature as the Key to Belonging in a Modern Industrial World

Up until the twentieth century, the people of Iceland lived on agriculture, with the addition of fishing in some places along the coast. This changed in modern Icelandic society. Now, most Icelanders live in Reykjavik, the capital of Iceland, and nearby towns.[7] But although the majority of Icelanders' livelihoods are no longer directly dependent on nature through farming or fishing, the vision of nature is still an integral part of Icelandic history and has set a clear mark on contemporary Icelandic society and attitudes. It derives not least from the fact that nature, through the use of its resources, is an important factor in the making and maintaining of modern techno-industrial society in Iceland.[8]

Ideas on economic progress and the use of natural resources, which characterized visions of nature in the Western world at the beginning of the twentieth century, were also firmly established in Iceland at that time. Iceland was then one of the poorest countries in Europe, and ahead was the daunting task of raising the nation from poverty and building a modern society. In the first half of the twentieth century, Icelandic modernizers looked for technical solutions to the persistent economic problems and poverty in their country, and politicians and intellectuals regarded the new century, with its technological opportunities, as the beginning of a new era in Icelandic history. The thinking was that Iceland would gain economic independence with modern industries and the prosperity necessary to sustain a sovereign and modern nation, since it was not destined to be a poor, undeveloped country. The plan was that in the near future it would take place among the rich and civilized nations.[9]

Iceland's hydropower capabilities were supposed to ensure this status. Icelandic natural resources, in this case the energy hidden in rivers and waterfalls, were to serve the modern needs of the nation. Thus, the glacial rivers caught the attention of progressive thinkers and nationalist intellectuals. They were for man to use. It was not only the prerogative of Icelanders to harness the energy of the Icelandic rivers but their duty and a national mission. Victories in the nation's struggle with Icelandic nature, with the damming of Icelandic rivers, became an important factor in identity formation in Iceland, underlining the modern and technologically advanced character of its society. Hydropower stations took on, therefore, symbolic value at the same time as they were built for economic purposes.[10]

Thus, the advent of the electrical age at the turn of the nineteenth century transformed the rivers into potential sources of wealth, and as time went on, turned many of them into actual sources of wealth as Iceland's basic energy resource. The industrial conquest of nature in twentieth-century Iceland not only had a very nationalistic tenor but was also pursued under the banner of progressivism and socio-economic policy.[11] This was clearly demonstrated in the celebrating speeches given by politicians, when the first big hydropower station was opened in 1970, harvesting the enormous power of Þjórsá, the longest glacial river in Iceland. Finally, Icelanders saw the dream of the progressive thinkers from about 1900 come true; Icelanders had managed to bend nature under their service instead of being her victims, as had been the case for centuries, said Jóhann Hafstein, the minister of industry.[12] Man's domination over nature was seen as a mission to serve progress and the well-being of the nation. It was for nature to provide and obey. This would be the only way for Icelanders to reach the level of prosperity that was thought to be their right as a modern, Western nation.[13]

In that vision there was little room for preserved and untouched nature. Such a thing was seen as a luxury that a modern nation could not afford if it wanted to prosper economically, and as an unrealistic goal in a modern world, as the engineer Jakob Björnsson put it in a newspaper article in 1969. It would be inevitable and unavoidable, he said, that with modernization, natural landscape would cease to exist. It had to be, and would be, transformed to serve man's needs. To think one could keep an area untouched if it had economic value was unrealistic. Cultural landscape would take over natural landscape in the largest parts of Iceland, he prophesized.[14] This somewhat technocratic vision was in harmony with the policy of the Icelandic authorities at the time. No river in Iceland should be spared if calculations showed it would be economical to harness it. This policy met limited opposition until after 1970. Until then, most Icelanders saw hydropower plants as a necessary and justifiable part of modern technological society.[15]

Dreams of harnessing the power of the glacial rivers in Iceland went hand in hand with dreams of large-scale energy-intensive industries, all of which were supposed to be powered by prospective hydropower plants. From the 1960 and up to the beginning of the twenty-first century, Icelandic entrepreneurs, politicians, and representatives of power companies attempted to persuade prospective foreign investors to invest in such industries in Iceland. In return, they were promised low-priced hydropower. This effort was rewarded, and aluminum smelters were built by international corporations Alusuisse (now Rio Tinto Alcan) and Century Aluminum in the south of Iceland, and Alcoa in the east.[16]

Victories in the nation's struggle with nature and efforts to bring it under the control of man with the damming of the rivers became an important

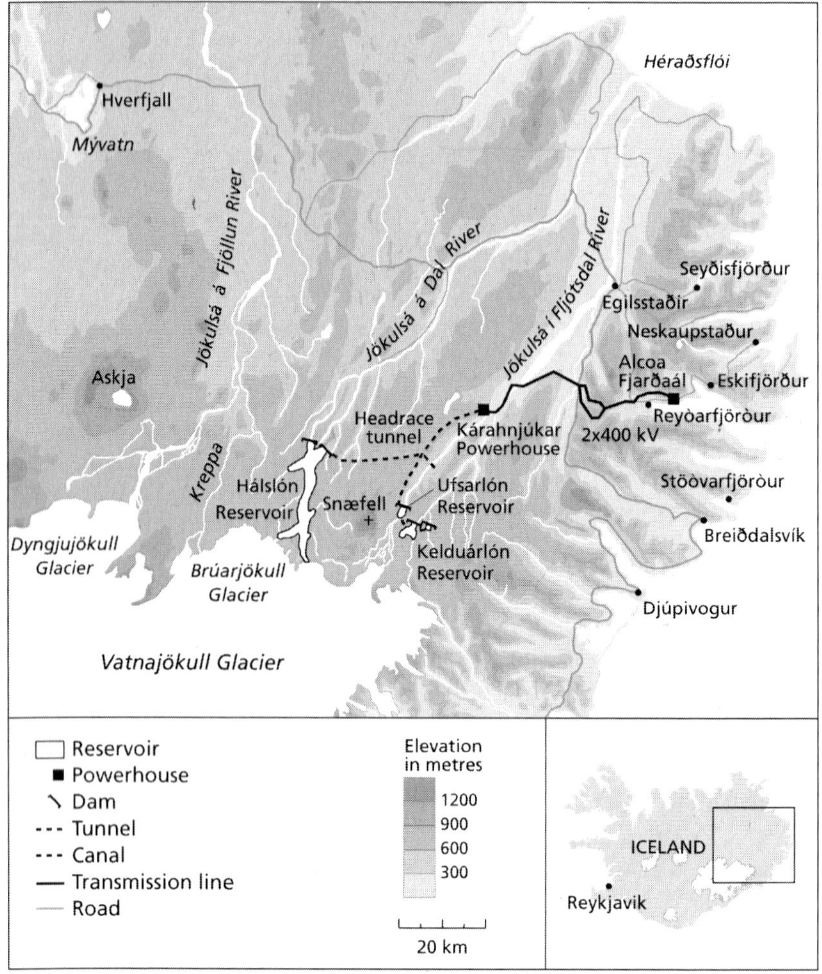

Map 11.1 The Kárahnjúkar hydropower project area

factor in identity formation in twentieth-century Iceland, underlining the modern and technologically advanced character of Icelandic society. The latest and largest project in that area, the Kárahnjúkar hydropower station in Iceland's eastern highlands (Map 11.1), can be seen as a spectacular symbol of that policy, with its almost 200-metre-high and 730-metre-long dam in Iceland's widest and deepest canyon. A heavy-streaming glacial river is blocked on its way to the sea and its stream diverted by tunnels over to the next valley, where its power is harnessed in the power station. From there it falls toward the sea, joined with the flow of another river, causing changes in that river's ecosystem. Another glacial river is also dammed up in the highlands, its flow directed to the power station by canals and tunnels.[17]

A construction of this magnitude has, of course, a wide-ranging impact on the environment. It changes the waterways of two big glacial rivers and causes the disappearance of many smaller rivers in the area. The creation of a twenty-five-kilometre-long reservoir submerges a fifty-seven-square-kilometre area of land. Smaller reservoirs, dams, roads, and canals also change the landscape. In sum, the ecosystem of the entire area is affected, from the highlands down to the coast and into the sea.

The Kárahnjúkar hydropower station is by far the largest power station in Iceland. It produces 690 megawatts, not for general public use but rather to provide energy for the aluminum smelter, which was built in a town in one of the fjords on the eastern coast of Iceland. The smelter is owned by the international company Alcoa. The Kárahnjúkar project was firmly supported by the Icelandic government and initiated by it as a part of its energy and heavy-industry policy. The Icelandic government and the developer promoting the project, the state-owned National Power Company of Iceland (Landsvirkjun), acknowledged the damage and changes to the regional ecosystem and landscape the power plant caused but justified them as necessary sacrifices for the sake of economic benefits and increased employment in the eastern region.[18]

Conflicting Views on Nature

Through the twentieth century, attitudes in Iceland to the rivers and nature in general were driven by strong anthropocentrism, and it still dominated governmental policy and public opinion at the outset of the twenty-first century.[19] Until the 1980s, the pressure was on energy companies to use the power of the rivers to produce electricity for homes and local industries, and from the 1980s, to produce it for heavy industrial use, such as aluminum smelting. This deeply marked the history of nature preservation in Iceland, as its standing was very weak vis-à-vis the government's energy policy and the demands of energy companies to harness rivers for hydropower production. The focus of nature preservation in Iceland has generally been on protecting the landscape on the grounds of nature romanticism and preservationism, on protecting landscape such as waterfalls and remote wilderness, as well as ecosystems. Many of these ecosystems are in the areas where the rivers flow, both in the lowlands and highlands. In the second half of the twentieth century, nature conservationists did win a few important victories in their hard battle for the protection of rivers and areas with special ecosystems.[20] Still, on the whole, where interests of energy projects clashed with the views of those who fought for the protection of nature, the conservationists tended to lose.[21]

Debates on hydropower plants set their mark on life in Iceland during the last quarter of the twentieth century and at the beginning of the twenty-first. These disagreements, flaring up after 1970, are rooted in the ideological heritage of the nineteenth century, which had formed Icelandic ideas on nature around the turn of that century. In the first half of the twentieth century only a few critical voices on the uncompromising policy of exploiting the natural resources in Iceland could be heard. But the opposition gained mass support in the years around 2000. International concerns about environmental issues began to influence views on nature in Iceland in about 1970 and have done so up to the present day, mixing with older nineteenth-century romantic views of nature based on aesthetic values. The romantic view on nature protection, influenced by nationalism and patriotism, emphasized the link between the land itself and the nation inhabiting it. Part of this ideology is the glorification of the natural beauty of Iceland; perceived beauty enhances the value of natural phenomena and sites.[22] In this picture, the waterfall has a special place.

The Sublime Waterfall – An Endangered Natural Phenomenon

Iceland is a land of many waterfalls, thanks to its landscape and climate, and they are often considered to represent the country's natural beauty (Figure 11.1). Waterfalls have been idealized in Icelandic literature since the romantic period of the nineteenth century in connection with ideas about nature's picturesque or sublime beauty, and even more so in the neo-romanticism period, during the first decades after 1900. This admiration and canonization of waterfalls in nationalistic nature romanticism and on aesthetic grounds, as symbols of Iceland's natural beauty, have been deeply embedded in Icelandic literary culture, arts, and mentality.[23]

The waterfall's place in the Icelandic attitude toward nature was clearly reflected in the fight against the harnessing of the big glacial rivers in the eastern highlands in the years around 2000. The protest's positioning aligns with romantic portrayals of waterfalls in Iceland. It was emphasized how each waterfall is a part of its surrounding landscape and yet has something so individualistic and sublime in its quality, shape, sound, and size that it is a special phenomenon standing out in the landscape.[24] This was, for example, reflected in the words and work of the Icelandic artist Rúrí, who was

Figure 11.1 The Faxi waterfall, which disappeared after the construction of the Kárahnjúkar dam. Faxi was twenty metres high and located in one of the two glacial rivers dammed because of the project. It was among the higher waterfalls that were lost in the Kárahnjúkar dam area. Photo by the author.

among those criticizing the destruction of the waterfalls in the eastern highlands. Waterfalls are a recurring theme in her multimedia art, and through her pieces she has demonstrated how the waterfalls in Iceland have their own characteristics. Some are light and joyful. Others are heavy and even murky from the tremendous power that they have. Thus, it is easy to become fascinated by the sound and shape of a waterfall, by its depth and dimensions, she said in 2008 when asked to describe the effect waterfalls have on her. Although Rúrí's work had a critical agenda, her purpose was to raise awareness of the fact that many waterfalls in Iceland were endangered because of the then upcoming grand-scale hydropower production plans.[25]

A powerful waterfall might give the expression that it has always been where it is. But it is not so, as waterfalls are neither unchanging nor eternal. They change over time for natural reasons. They gradually retreat or dig themselves deeper into the rock down under, inch by inch, as the water demolishes the rock, and change in shape and size, depending on the flow of the river and how it digs its channel; they disappear if the river dries up or its channel is altered. Nature in Iceland has created waterfalls and made them disappear again through its geological history. But the most rapid change has been caused by man – and only in the last few decades. The Icelandic geologist Sigurður Þórarinsson warned at the end of the 1970s that man would be the greatest danger to waterfalls in Iceland because of plans to harness all the largest rivers, and many of the smaller ones as well, for hydropower production.[26]

His words came true. Since the 1970s, many waterfalls have disappeared from the Icelandic landscape for good, and the number is even greater if those lost over the previous hundred years, when Icelanders began to harness rivers for hydropower production, are added. According to governmental energy policy in about 1980, most of the large and medium-sized waterfalls were to disappear altogether or be decreased in size because of dams for hydropower plants. None of the bigger waterfalls was excluded in these plans, not even those most highly valued and adored for their size and sublime character and which are among the country's greatest tourist attractions – for example, Gullfoss and Dettifoss.[27] Gullfoss was protected by law in 1979, and Dettifoss in 1996, as a result of pressure from nature conservationists. Still, the heavy-streaming glacial rivers in which they fall were not also protected at the same time and were later included in power project plans, until public opinion against those plans put them on hold and they were set aside.[28]

The "harness it all" policy of exploiting the rivers as energy sources has been heavily criticized since about 1970. This opposition, fought under the banner of nature preservation, has forced energy authorities to set some of the power project plans on hold and drop others. They have done so in part because there has not been enough demand for the energy, and in part in response to appeals for the protection of Iceland's most beautiful and best-loved waterfalls, which became louder and more widespread in the wake of increasing environmental awareness in the country from about 1970 onward.

Nevertheless, the battle for preserving the waterfalls in the eastern highlands was lost in 2002, when the Icelandic Parliament passed a bill that granted permission for the construction of the Kárahnjúkar hydropower station. As a consequence, about seventy waterfalls disappeared, some of them up to thirty metres high. By way of comparison, the most famous Icelandic waterfalls, Gullfoss and Dettifoss, are thirty-two metres and forty-four metres, respectively. Never before in the history of Iceland had one hydropower plant destroyed so many waterfalls. This was a turning point in the Icelandic vision of nature. At the beginning of the twentieth century, there were plenty of waterfalls in Iceland, and people saw the country as such – as a land of plentiful waterfalls. But a hundred years later, the waterfall was seen as an endangered phenomenon in Icelandic nature.[29]

Beautiful Landscape – Lost Landscape

Nature is important for Icelanders in the twenty-first century in their efforts to create an image of Iceland as a world apart, as a unique place in the world. The image of Iceland as the land of ice and fire is promoted, highlighting the island's glaciers and volcanoes, as is the image of Iceland as a country of clean and sublime nature, and with the largest wilderness area in Europe. One of the most emphasized parts of Iceland in this image making is the highlands, promoted to the world as a vast, remote, unspoiled wilderness, worthy to see and experience as one of the few such spots left in the world.[30]

The strong anthropocentric idea that Icelanders have an unquestionable right to exploit their country's nature, which dominated energy policy for over a hundred years, has created a deep controversy. It conflicts with the views of many Icelanders on the meaning of nature for Icelanders in today's modern world. Recent research on the views on nature among Icelanders shows that they most appreciate a landscape that has not, or to a limited extent, been altered by man – a natural landscape – and that such a landscape is seen to be what first and foremost symbolizes Iceland today.[31]

This high regard for the natural landscape was reflected in the fight from 1999 to 2006 against the hydropower project in the eastern highlands. The area there that was to be exploited for hydropower production became a symbol of unspoiled nature – of Icelandic wilderness. It also became a symbol for the danger such nature faces, not only in Iceland but all over the world today because of industrialization, engineering, and technocratic opinions enforced by the power of technology and machinery. Thus, the concepts of untouched and unspoiled were central in the rhetoric of the campaign to save the eastern highlands from being sacrificed for hydropower production. This was an idealization of wilderness, its natural sites untouched by man. The eastern highlands were idealized as vast, remote, sublime, and untouched – beyond the touch of human activity. Any man-made, artificial landscape in this area that would replace the natural one could not, according to this view, be considered beautiful.[32]

The concept of untouched or unspoiled is problematic here, of course, as is the case in so many other parts of the world when it comes to defining changes in nature caused by humans. But the purpose of using these concepts in the fight to protect the Icelandic highlands was to idealize their nature according to modern wilderness romantic views focusing on aesthetic characteristics of the landscape, rather than looking at it from an environmental perspective – for example, as an area that suffers serious erosion in large parts.[33]

The highlands are the pearl of Icelandic nature in the eyes of those who fought against the construction of the Kárahnjúkar hydropower station.[34] Among the environmental arguments used in the conflict, ideas about nature as having a higher purpose for man to discover, a kind of romantic metaphysic, and animism cropped up. The highlands were viewed not only as a place characterized by scenic landscape and shaped by the forces of nature but also as a world of spirits, with its anima, its own purpose, will, and wishes. This is evidenced by the use of Hákon Aðalsteinsson's poem "Invocation" on protest posters, making it one of the dominant rhetorical symbols used in the fight against the power project. In his poem, Hákon Aðalsteinsson calls upon the spirit of the highlands to give him the strength to guard its treasures, its landscape, flora, and fauna. Thus, the highlands were seen as more than a physical world. They were also a world of a living spirit and with a voice of their own, uniting the smallest flower and the biggest mountain.[35] The poem praises the highlands and reflects the author's strong ties to them and his desire to protect them. Aðalsteinsson, a farmer, poet, and guide for reindeer hunters in the eastern highlands, was one of the local people in eastern Iceland who fought hard against the plans to harness the glacial rivers there. The appropriation of this poem in such a complex political battle, where environmental concerns and industrial and socio-economic interests clashed, demonstrates that it is easy for many Icelanders to spiritualize nature. It also demonstrates the fact that the idea that nature exists partially outside the physical, rationally explainable world has considerable support among the Icelandic people.

Taking a big slice of land in the eastern highlands and turning its natural landscape into industrial landscape meant destroying invaluable treasures in the eyes of those who admire and love the Icelandic wilderness. Þorvaldur Þorsteinsson, an Icelandic artist and writer, compared the destruction of Iceland's wilderness caused by the Kárahnjúkar hydropower station to that caused if someone were to walk into an art gallery and destroy some of the masterpieces of the old Icelandic landscape painters. It is a complete contradiction, he said, that Icelanders adore and carefully preserve the paintings of Icelandic landscape but destroy the landscape itself – the original, the art of nature.[36]

For many participants in this debate, the site was sacred, in a way. In their minds they conflated Christian ideas about nature as God's creation with ideas stemming from romanticism and its idealization of nature as a sublime phenomenon and meaningful to man as such. The beauty of the nature in this area was seen as being highly valuable, not in a materialistic sense but for man's mental, emotional, and spiritual well-being and growth.[37] Hence, some of the arguments for its preservation carried a religious tone. The highlands are a holy place for Icelanders, wrote the Icelandic Lutheran minister Gunnar Kristjánsson, and he asked if it would not be more sensible for Icelanders to treasure the highlands in their unspoiled state, and to learn to enjoy its extraordinary beauty, than to hand them over to foreign industrial corporations at a low price.[38]

The opponents of the Kárahnjúkar project pointed out that it was a huge contradiction that Icelanders should be so preoccupied with the beauty of Icelandic nature while willing to spoil a vast and remote wilderness. Consequently, the world reputation of Icelanders as keepers of untouched nature would suffer because they were failing in that role. This was a big mistake, said Icelandic writer and environmentalist Andri Snær Magnason. He wondered whether Icelanders, inhabitants of a nation that ideologically and economically thrives on the concept of the beauty of Icelandic nature, understood beauty at all. He was afraid that the world would start to separate Iceland from Icelanders: Iceland would be a symbol of nature's beauty, whereas an Icelander would be a symbol of what threatens it and doesn't understand it.[39]

This view of Iceland's natural environment as beautiful and unique, in the spirit of the romantic view of nature, is also reflected in the words of Iceland's best-known celebrity, the musician Björk Guðmundsdóttir. She has given much of her time and effort over the last few years fighting for nature preservation in Iceland. In June 2008, she gave an interview in Iceland to promote her upcoming concert in Reykjavik, called "Nature." The goal of the concert was to raise awareness among Icelanders of the importance of protecting Iceland's natural environment. Iceland is known around the world, Björk said, despite it being just a small island, and this is because of its nature. Many people consider Iceland as some kind of remote paradise, she added, because it has a certain primordial power, and represents something pristine, pure, and genuine. Because of this, many people want to come to Iceland, to experience the magic of Icelandic nature. It would be terrible if Icelanders destroyed this aspect of the country, she concluded.[40]

Thus, the battle against the Kárahnjúkar hydropower project was on the one hand a hard struggle against the project itself and on the other hand a long memorial ceremony for the nature that was to disappear under reservoirs and be transformed by dams, roads, and other such constructions. The sublime beauty of nature in the eastern highlands was the main theme in this prolonged commemoration during the protests from 1999 to 2006.[41]

The Nation as the Guardian of the Natural Heritage

In the struggle against the Kárahnjúkar hydropower project, the main arguments were in line with weak anthropocentric environmental policy, which emphasizes human interests in preserving the environment, in terms of the adoration of certain landscapes, romantic attachment to the wilderness, and the culture of outdoor life. These arguments were mixed with ecocentric reasoning, focusing on people's responsibility toward nature and the importance of not disrupting the ecosystem or the delicate balance of the various forces of nature. It was also argued that preserving the country's ecosystems as intact as possible was the foundation for cultural and economic prosperity.[42]

Although the protests against the hydropower project in the eastern highlands were driven by a growing environmental awareness, there was a strong tendency among the protesters to use nationalistic rhetoric and symbols from earlier times, particularly the cultural-political heritage of the nineteenth and early twentieth centuries. Among the general public, the fight against the big hydropower station was primarily fought under nationalist banners rather than that of contemporary environmentalism and green politics. The rhetoric and symbols of the nationalist era before 1944 were brought back to life and given a new purpose as an instrument in the fight to protect the highlands against irreparable damage. At protest meetings, people quoted in their speeches or chanted the nationalistic natural poetry of the nineteenth and early twentieth centuries, to express their love of Icelandic nature and their will to protect it. Symbolic demonstrations to protest the hydropower dam often had nationalistic undertones too.[43] By this the protesters wanted to emphasize their opinion that spoiling nature in the highlands was equivalent to desecrating national symbols. For example, the statue of Jón Sigurðsson, the Icelandic nineteenth-century national hero, was wrapped in aluminum paper to protest, in a token manner, the plan to destroy vast wilderness in the eastern highlands for the only purpose of harnessing energy for a foreign-owned aluminum smelter.[44]

Another symbolic act demonstrating the close relationship between the nation and nature in Iceland, which also referred to national symbols, was the laying of stones on the spot where the reservoir would be formed if the dam was to be built. One word from the first verse of the Icelandic national anthem was etched on each stone, and the stones laid on the surface of the ground so that one could read the entire verse by walking through the area, which was to be sunk under water as part of the proposed reservoir. The linking of nature and the revered national symbols was meant to demonstrate that to sink this wilderness under water amounted to an attack on the nation itself.[45]

The nationalist tone of the campaign was also motivated by the fact that the Kárahnjúkar hydropower station was to be constructed solely for the purpose of producing energy for a foreign-owned aluminum smelter. Icelandic authorities were harshly criticized for taking a valuable natural area away from Icelanders in order to lure foreign corporations into building aluminum smelters in the country.[46]

To emphasize the significance of the highlands further, the opponents of this massive hydropower project placed the nature of the highlands at the same level as the medieval manuscripts stored in the Arnamagnæan Institute in Reykjavik. These cultural treasures are regarded as the most valuable national symbols in Iceland, representing Iceland's only major contribution to world culture. Admiration for the medieval Icelandic sagas was crucial in constructing early Icelandic national identity, through the pride of owning something considered precious and irreplaceable not only in the eyes of the nation itself but also in those of many foreign observers.[47] According to the protesters, Icelandic nature had written its own scripts in the highlands in a unique and magnificent way. These were masterpieces, which no man-made construction could ever imitate or recreate. Destruction of these natural treasures was equivalent to destroying the most valuable of Icelandic cultural treasures, the medieval manuscripts, they claimed. Those who fought for the construction of a power station in the highlands were fuelling the Icelandic economy with nature's manuscripts.[48]

Moreover, as mentioned, not only nature but also the image of Iceland as the land of unspoiled wilderness, the largest in Europe, would be damaged. Icelanders had the duty to guard their natural heritage just as they did the cultural heritage. This duty to protect Icelandic nature included Icelanders' international responsibility. Icelandic nature is not the private property of the Icelandic nation, the Icelandic journalist and filmmaker Ómar Ragnarsson wrote during his own campaign to save the eastern highlands in the east from disappearing under a big reservoir. Icelandic natural landscape, with its ecosystem, is a treasure belonging to all mankind, not only Icelanders, he argued.[49]

Another side of this story was the belief held by many Icelanders that the highlands are the common property of all Icelanders, and therefore that the public has a right to have a say on how it is allocated and exploited.[50] These ideas, which have their roots in the 1990s, were intertwined with dreams about preserving all the highlands. The notion was to create a huge nature reserve by protecting all of what was then left untouched of the highlands – that is, the vast area not yet changed by hydropower stations or other man-made constructions.[51] The environmentalist and writer Sigrún Helgadóttir was one of the proponents of these ideas, arguing that it was the collective right of all Icelanders to enjoy the area and the satisfaction that it provided as an untouched wilderness. She also thought it the duty of the entire Icelandic nation to protect the highlands against changes or damage to its landscape and ecosystems caused by human utilization.[52]

The strong relationship between nationalism in Iceland and ideas about nature can be explained by the country's cultural and political history. Icelandic political discourses have been steeped in cultural nationalist rhetoric, and therefore it is easy for most Icelanders to identify with the symbols of Icelandic nationalism. Thus, it seemed natural to use it as an instrument in the battle from 1999 to 2006 against damming the glacial rivers in the eastern highlands. Most Icelanders are brought up learning to appreciate landscape paintings and patriotic nature poetry, which emphasizes and idealizes the beauty of the country. Reading and learning nature poems in school was a part of an education policy for much of the twentieth century, and the twentieth-century naturalist painters, who made the sceneries and elements of Icelandic nature their main subject, were, and still are, highly treasured among Icelanders.

Belonging and Being Apart

There is an innate link between nationalistic values and the love of nature because the basic idea of nationalism centres on the essential relationship between a country – the homeland – and a nation. In the battle to protect the natural landscape of the Icelandic highlands, this was reflected in the views of those who saw Icelandic nature and the Icelandic nation as one, inseparable entity, thereby looking at the Icelander as a part of Icelandic nature and, vice versa, Icelandic nature as a part of the Icelander.[53] According to some conservationists, this meant that true loyalty to one's country implied that one's thoughts and deeds should aim at benefiting both the present and the future of the country and the nation, because the country and the nation were in essence two sides of the same coin. Furthermore, this merging of country and nation would put the moral obligation on the shoulders of every Icelander not to destroy the country's wilderness but to hand it over unspoiled to future generations. Such unspoiled nature was considered to be the only national heritage and a national treasure of any true value, a birthright of every Icelander.[54]

Thus, the battle to protect the highlands was not only about preserving landscape and ecosystems but also about attitudes to life and the meaning of being Icelandic; it was about protecting what was regarded as immeasurable treasures, touching the core of life and what it means to be an Icelander. This view is represented in the writings of Guðmundur Páll Ólafsson, an Icelandic biologist and writer.[55] The Icelandic philosopher Sigríður Þorgeirsdóttir also describes the highlands as the heart of Iceland. In her writings about the importance of protecting the highlands, she sees this in a spiritual context because of how much the identity of Icelanders is shaped by the closeness and coexistence with natural forces and the landscape. The highlands give Icelanders the strength they need to survive and prosper, she argues, and it has an important and unique ideological and economical value. Unspoiled nature is very important for Iceland's international image, and it is the pride of the nation, she writes.[56] Páll Skúlason, professor of philosophy at the University of Iceland and its rector from 1997 to 2005, also wrote about the meaning of the highlands for Icelanders as a nation. Icelanders have to think about how important the highlands are for the national consciousness, he has claimed, because Icelandic nature – in this case, the highlands – shapes the way Icelanders think more than they realize. If the highlands were to be sacrificed to the construction of hydropower stations, it would mean a fundamental change for Icelanders, who generally picture themselves as living in a country characterized by a wild nature that man cannot control.[57]

In the eyes of those who fought against the hydropower project in the east, the Icelandic highlands play a key role in national identity formation in Iceland today and in the image that Icelanders want to project of themselves to the world, as a nation of the North in close relationship with nature. Thus, the fight in the years around 2000 against the Kárahnjúkar hydropower project was very much about the issue of how to shape and formulate the image and identity of the Icelandic nation. The highlands were the centrepiece in that picture, and seen as a unique wilderness for its vastness, ecosystems, and landscapes created and continuously affected by fire and ice. Its uniqueness was thought to give the Icelandic nation a special place in the world of nations, since it is the owner and guardian of such a special nature, a wild area such that so many nations do not have any more, having lost it to resource exploitation, construction, and settlements. If Icelanders would let the highlands be spoiled, and thereby lose their natural treasure, the Icelandic nation would lose its uniqueness as well. Without the vast, unspoiled wildernesses of the highlands, Icelanders would be just like any other Western nation, living in any other industrial consumerist society.[58]

The debate on hydropower projects in Iceland for the last ten to fifteen years has divided the Icelandic nation into two groups. In essence, the conflict is not only about preservation of pristine nature versus pro-industrial policy concerning energy resources – it is a clash of irreconcilable values. On the one side are those who want to continue to use Icelandic nature according to the policy set in the early 1900s, combining the idea of progress and the exploitation of nature and its resources. On the other side are those who want to pursue post-industrial strategies in building up the economy and labour market in Iceland, in order to keep what is left of its wild nature intact.[59]

It is clear that Icelanders put great value in the meaning of being part of the North as a community of nations, with a certain kind of common identity, based on historical and cultural ties. The feeling of belonging to a supranational society within a certain hemisphere shortens geographical distances in a subjective way. It creates a sense of unity. Thus, the North is a cultural capital, and it builds bridges between nations in a constructive and positive way. But on the other hand, the idea of the North as nature, and within it Icelandic nature as something special in the world, distances the Icelandic nation from the rest of the world. This relationship with nature gives the sense of being apart, the feeling of not belonging to anything but one's country, and thereby defining one's place in the world. So, the meaning of the term "North" is diverse, even within the same country, as the case of Iceland shows.

NOTES

1 See M.C. Hall, D.K. Müller, and J. Saarinen, *Nordic Tourism: Issues and Cases* (Toronto: Channel View Publications, 2009).

2 Guðmundur Hálfdanarson, *Íslenska Þjóðríkið: Uppruni og endimörk* (Reykjavik: Hið íslenska bókmenntafélag og Reykjavíkur Akademían, 2001). See also Birgir Hermannsson, *Understanding Nationalism: Studies in Icelandic Nationalism, 1800-2000* (Stockholm: Stockholm University, 2005), 287-99.

3 Guðmundur Hálfdanarson, "The Nordic Area: From Competition to Cooperation," in *Empires and States in European Perspective*, ed. Steven Ellis (Pisa: Edizioni Plus, 2002), 83-93.

4 On the issue of language for Icelanders in Nordic cooperation today, see, for example, Katrín Jakobsdóttir, "Norðurlöndin eiga samleið," *Skíma: Vettvangur móðurmálskennara*, http://www.modurmal.is/. See also *Norðurlandamálin með rótum og fótum*, Nord 2004 (12), ed. I.S. Sletten (Copenhagen: Norræna ráðherraráðið, 2005).

5 See Stefán Ólafsson, *Íslenska leiðin: Almannatryggingar og velferð í fjölþjóðlegum samanburði* (Reykjavik: Háskólaútgáfan, 1999); Stefán Ólafsson, "Þróun velferðarríkisins," in *Íslensk þjóðfélagsþróun, 1880-1990: Ritgerðir*, ed. Guðmundur Hálfdanarson and Svanur Kristjánsson (Reykjavik: Félagsvísindastofnun Háskóla Íslands and Sagnfræðistofnun Háskóla Íslands, 1993), 399-430. See also Svanur Kristjánsson, "Stjórnmálaflokkar, ríkisvald og samfélag, 1959-1990," in Hálfdanarson and Svanur Kristjánsson, *Íslensk þjóðfélagsþróun*, 355 98.

6 Guðni Th. Jóhannesson, *Hrunið: Ísland á barmi gjaldþrots og upplausnar* (Reykjavik: JPV útgáfa, 2009).

7 There are today approximately 300,000 Icelanders: about 191,000 live in Reykjavik and nearby towns, and about 109,000 in other parts of the country. For statistics in Iceland, see Statistics Iceland, http://www.statice.is/.

8 Hálfdanarson, "The Nordic Area," 99-215.

9 Ólafur Ásgeirsson, *Iðnbylting hugarfarsins: Átök um atvinnuþróun á Íslandi, 1900-1940* (Reykjavik: Menningarsjóður, 1998).

10 Unnur Birna Karlsdóttir, *Þar sem fossarnir falla: Náttúrusýn og nýting fallvatna á Íslandi, 1900-2008* (Reykjavik: Hið íslenska bókmenntafélag, 2010), 23-74.

11 Ibid., 85-86, 203-5.

12 Jóhann Hafstein, "Í kapphlaupi við þann sem ekki mæðist – tímann sjálfan," *Morgunblaðið*, blað II, 5 May 1970, 2.

13 Karlsdóttir, *Þar sem fossarnir falla*, 86.

14 Jakob Björnsson, "Almennt um náttúruvernd á Íslandi í framtíðinni," *Verkamaðurinn*, 3 October 1969, 2.

15 Karlsdóttir, *Þar sem fossarnir falla*, 23-84.

16 Ibid., 85-88, 147-52, 190-201.

17 *Kárahnjúkar: Hydroelectric Project Iceland* (Reykjavik: Landsvirkjun, 2009).

18 Karlsdóttir, *Þar sem fossarnir falla*, 158-76.

19 Ibid., 206.

20 Ibid., 83-84, 89-113, 115-41.

21 Ibid., 214-15.

22 Hálfdanarson, *Íslenska Þjóðríkið*, 191-216.

23 Karlsdóttir, *Þar sem fossarnir falla,* 210-11. See also Guðmundur Hálfdanarson, "'Hver á sér fegra föðurland': Staða náttúrunnar í íslenskri þjóðernisvitund," *Skírnir* 173, 2 (1999): 304-36, and Þorvarður Árnason, "Knỳ minn huga, gnýr: Dettifoss með augum Einars Benediktssonar," in *Heimur ljóðsins,* ed. Ástráður Eysteinsson et al. (Reykjavik: Bókmenntafræðistofnun Háskóla Íslands, 2005), 317-34.

24 Karlsdóttir, *Þar sem fossarnir falla,* 29-35, 41-44, 187-90.

25 An interview on the program *Listin og Landafræðin* with the artist Rúrí on Icelandic National Radio (22 June 2008). Her art pieces on waterfalls are *Archive Endangered Waters* (2003), *Water Vocal Endangered* (2007), *Silence of the Waterfall* (2007), and *Flooding* (2007).

26 Sigurður Þórarinsson, *Fossar á Íslandi* (Reykjavik: Náttúruverndarráð, 1978), 7-8.

27 Jakob Björnsson and Haukur Tómasson, *Umsögn Orkustofnunar um tvö fjölrit Náttúruverndarráðs: "Fossar á Íslandi" og "Vatnavernd"* (Reykjavik: Orkustofnun, 1979), 1-8.

28 Karlsdóttir, *Þar sem fossarnir falla,* 77-84, 158, 192, 196.

29 Ibid., 162, 188-89. See also Pétur Ingólfsson and Gerður Jensdóttir, *Áhrif Kárahnjúkavirkjunar á fossa,* (Reykjavik: Landsvirkjun, 2008).

30 An example of an Icelandic information website that promotes this image of Iceland is *Inspired by Iceland,* at http://www.sagenhaftes-island.is/. The website's ideal image of Iceland as an exotic world of ice and fire, pure nature and pristine wilderness, is somewhat biased, of course, no matter what one thinks of the nature it idealizes, although the purpose is to promote a positive image of Iceland and promote it as a tourist attraction.

31 Þorvarður Árnason, "Views of Nature in Iceland: A Comparative Approach," in *Views of Nature and Environmental Concern in Iceland* (PhD diss., Linköping University, 2005), 103-21.

32 Karlsdóttir, *Þar sem fossarnir falla,* 176-77.

33 On the wilderness ideology, see, for example, William Cronon, "The Trouble with Wilderness: Or, Getting Back to the Wrong Nature," in *Uncommon Ground: Toward Reinventing Nature,* ed. William Cronon (New York: Norton, 1995), 69-90. See also Greg Garrard, *Ecocriticism,* 2nd ed. (London: Routledge, 2005), 59-84. On erosion in Iceland, see, for example, Ólafur Arnalds et al., *Soil Erosion in Iceland* (Reykjavik: Agricultural Research Institute, 2001).

34 On this, see Rannveig Magnúsdóttir, "Ísland og stóriðja," *Morgunblaðið,* 9 November 2002; Jón Á. Kalmansson, "Tilfinningar, skynsemi og stórvirkjanir," *Lesbók Morgunblaðsins,* 15 March 2003; and "Draumalandið sem hverfur," *Fréttablaðið,* 8 July 2006.

35 Hákon Aðalsteinsson, "Ákall," in *Imbra* (Akranes, Iceland: Hörpuútgáfan, 2002), 13. The poem's title in English as "Invocation" is my translation.

36 Þorvaldur Þorsteinsson, "Þegar eftirmyndin verður frummyndinni yfirsterkari" (paper presented at a meeting of Náttúruvaktin, Reykjavik, 25 January 2005), http://www.natturuvaktin.com/.

37 Karlsdóttir, *Þar sem fossarnir falla,* 186. See also Ómar Ragnarsson, *In Memoriam?* (Reykjavik: Hugmyndaflug, 2003).

38 Gunnar Kristjánsson, "Stríð streymir Jökla," *Morgunblaðið,* 29 August 2006.

39 Andri Snær Magnason, *Draumalandið: Sjálfshjálparbók handa hræddri þjóð* (Reykjavik: Mál og menning, 2006), 258.
40 An interview by Ríkissjónvarpið with Björk Guðmundsdóttir on the evening news on Icelandic National Television, 5 June 2008.
41 The protests were acted out in various forms: public demonstrations, meetings, art performances, publications of books and newspaper articles, programs in the media, concerts, political debates within and outside the parliament, banners, manifestos, and slogans.
42 Karlsdóttir, *Þar sem fossarnir falla,* 172-78, 208-14.
43 Hálfdanarson, *Íslenska þjóðríkið,* 211-15. See also Ann Brydon, "Náttúran, mótmæli og nútíminn: Náttúrusýn Íslendinga – með augum gestsins," *Tímarit Máls og menningar* 62, 2 (2001): 45-49.
44 "Pökkuðu styttunni af Jóni inn í álpappír," *Morgunblaðið,* 7 September 2002.
45 "Þjóðsöngurinn til verndar hálendinu," *Morgunblaðið,* 7 September 1999.
46 This view was demonstrated in many articles written against plans of damming the glacial rivers in eastern Iceland. See, for example, Hákon Aðalsteinsson, "Hvert stefnum við í umgengni við land og þjóð?" *Morgunblaðið,* 21 October 1999; Guðbergur Bergsson, "Það að fórna náttúrunni," *DV,* 15 December 1999; Elísabet Jökulsdóttir, "Aðeins eitt land," *Morgunblaðið,* 11 October 2002; and Reynir Harðarson, "Sigur viljans," *Morgunblaðið,* 2 June 2006.
47 Hálfdanarson, *Íslenska þjóðríkið,* 198.
48 Guðmundur Páll Ólafsson, "Að eyða handritum," *Dagur,* 4 December 1998.
49 Ómar Ragnarsson, *Íslands þúsund ár* (Reykjavik: Hugmyndaflug ehf., 2006), 1.
50 Hálfdanarson, *Íslenska þjóðríkið,* 208-15.
51 Gísli Sigurðsson, "Framtíðarsýn á Fjöllum," *Lesbók Morgunblaðsins,* 22 June 1996; and Jónas Kristjánsson, "Víðerni allra landsmanna," *DV,* 15 August 1997.
52 Sigrún Helgadóttir, "Íbúðin Ísland," *DV,* 8 November 1990.
53 Guðmundur Hálfdanarson and Unnur Birna Karlsdóttir, "Náttúrusýn og nýting fallvatna," in *Landsvirkjun, 1965-2005: Fyrirtækið og umhverfi Þess,* ed. Sigrún Pálsdóttir (Reykjavik: Landsvirkjun, 2005), 165-99.
54 Guðmundur Páll Ólafsson, "Grát fóstra mín," *Morgunblaðið,* B, 19 January 1997.
55 Ólafsson, "Að eyða handritum," *Dagur,* 4 December 1998.
56 Sigríður Þorgeirsdóttir, "Heild sem gerir mann heilan: Um siðfræði náttúrufegurðar," in *Hugsað með Páli: Ritgerðir til heiðurs Páli Skúlasyni sextugum,* ed. Róbert H. Haraldsson, Salvör Nordal, and Vilhjálmur Árnason (Reykjavik: Háskólaútgáfan, 2005). 163-79.
57 "Gildi öræfanna í íslenskri þjóðarvitund," *Morgunblaðið,* B, 15 March 1996.
58 Karlsdóttir, *Þar sem fossarnir falla,* 213-14.
59 Ibid., 157-81.

EPILOGUE

The Networked North

Thinking about the Past, Present, and Future of Environmental Histories of the North

FINN ARNE JØRGENSEN

Reading the chapters in this collection of environmental histories of the North activates many parts of both my personal and academic identities. First, the chapters continually remind me that I am a Northerner. I grew up in a small Norwegian town, north of the Arctic Circle, on a skinny stretch of land between a fjord and a mountain. I remember that particular blue light that you get only in the northern winter and the occasional aurora borealis shimmering across a dark, star-filled sky, contrasted by the near-endless bright summer days. The rain and the darkness have mostly faded into the past, as the everyday tends to do. Since then, I have studied and worked in several places in the North – all university towns, with everything that follows. None was an isolated place: people came from all over the world to live, work, and study there; a short flight would take you to a major airport hub, opening up the world for travel; wide selections of goods and services were available. Perhaps it is not so surprising that these places are so connected; the people living there are acutely aware of the need to seek connections with the rest of the world. As a result, we end up with highly dynamic places, places that have shaped my outlook on and experience of the world. Although I have also lived in places that, distinguished by other climates, could not be called part of the North, certain climate conditions and landscapes have become the base from which I evaluate and relate to other places. Environmental history often deals with the making of place, but this is perhaps even more a feeling, a mental and emotional comparison between an imagined landscape that feels like home and whatever place I happen to be in.

At the same time, my academic identity and training have blended with this personal domestication of the North as a place and an outlook. I am a historian of technology and environment, with a background in science and technology studies and a special interest in the history and sociology of consumption. Rather than consider particular arrangements of nature and people as "natural" and self-evident, I see them as the result of historical processes, cultural choices, and technological networks. I think about how these places on the periphery are connected to the rest of the world through a multi-directional flow of people, information, resources, and commodities. The North is also a result of layers of narratives; fictional, scientific, mythological, and so on. Like any other place, the North is really a multitude of places, utterly ordinary and completely special.

Finally, I am also an organizer, frequently thinking about how to bring people together in networks and events of different configurations, not just to share and mobilize existing knowledge but also to generate new ways of thinking about the past, present, and future of the world around us. Just like space and place matter in defining the North, so do people. In 2009, I launched the Nordic Environmental History Network (NEHN), with generous funding from Nordforsk, the research branch of the Nordic Council of Ministers, to create a meeting place for scholars sharing this interest in environmental history. The field has a relatively long history in the Nordic countries, but very few institutions have been able to develop a critical mass of faculty and students researching, teaching, and learning about environmental history. The question of how to develop a vibrant and sustainable community in such a geographically wide-ranging setting is not a trivial one. I believe this volume and the process behind it prompts us to pay more attention to the networks of scholars doing environmental history, and thus the mechanisms through which we communicate, collaborate, and challenge each other. The network grant allowed a group of us, ranging from master students to senior professors, located in institutions of all sizes across the Nordic countries, to meet regularly, in Trondheim, Roskilde, Helsinki, Tromsø, and Stockholm (Map E.1). The sustained conversations and connections that arose from the broad mix of competencies and experiences have really helped create a sense of transnational community among environmental historians that I would argue did not exist in the same sense before. I think the last years' major environmental history conferences in

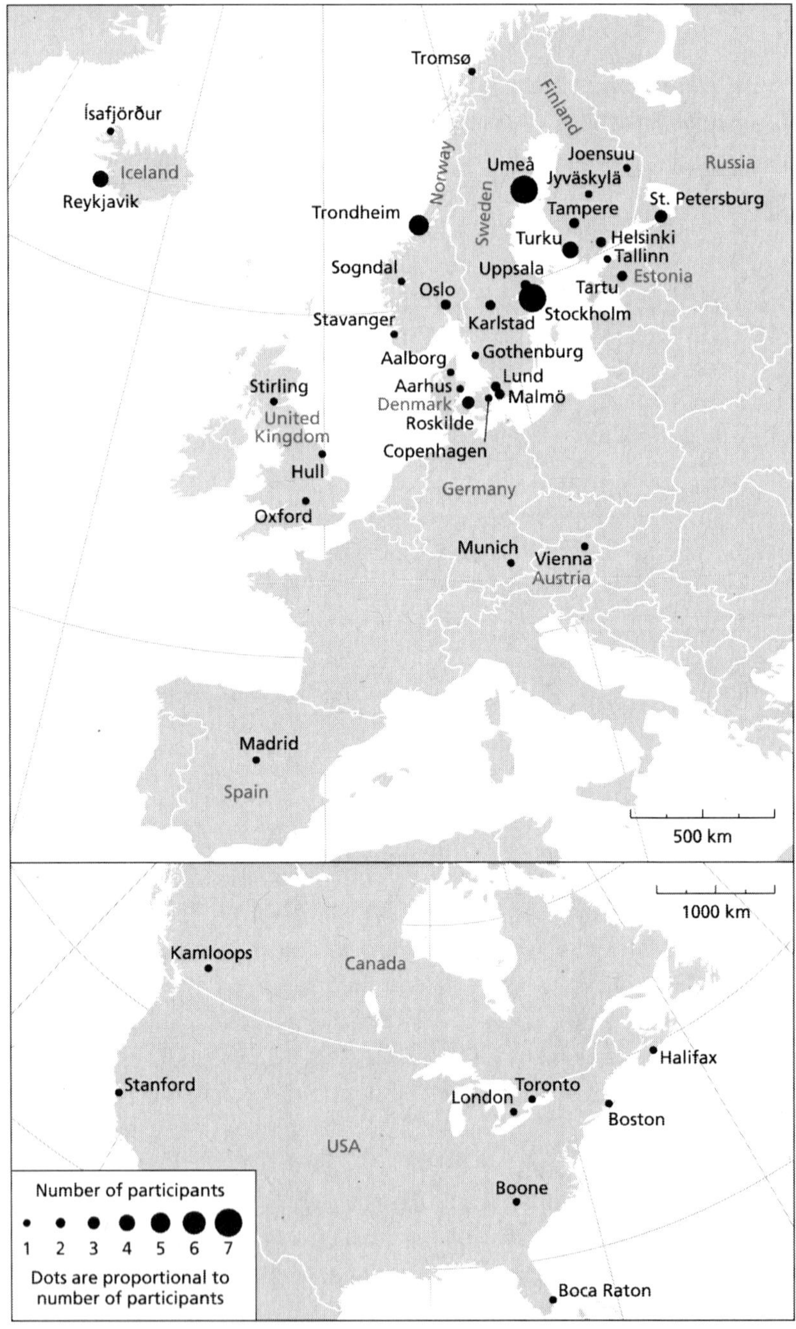

Map E.1 Provenances of participants in the Nordic Environmental History Network meetings, 2009-2012

the Nordic countries – the 2005 Nordic Environmental History Conference in Turku, Finland; the 2009 World Congress in Environmental History in Copenhagen, Denmark; and the 2011 European Society for Environmental History meeting, also in Turku – have strongly contributed to making the Nordic countries a major international hub in the environmental history community.[1]

This volume grows out of a meeting I arranged in Stockholm together with Sverker Sörlin in 2010, and demonstrates well the many-layered complexity of the North. In organizing the meeting that first brought together the contributors to this book, we gathered a truly circumpolar selection of scholars – not just from the Nordic countries but also from Canada, the United States, the United Kingdom, and Russia – to bring a diversity of perspectives to the table. During two October days in Stockholm, and over countless subsequent email exchanges, we jointly explored the environmental histories of the North that form the core of this book. Although national and regional boundaries can be arbitrary, particularly from an environmental history point of view, such networks provide a good base from which to work, especially if the participants strive for open boundaries and active collaboration with others. I believe environmental historians, and scholars in general, should strive for transnational research collaboration with each other to avoid limiting the scope of environmental history research to national or regional contexts. At the same time, this provides an opportunity to re-examine the formation of such contexts.[2]

The ways in which we as environmental historians work together do matter, and the regional networks such as NEHN, the Network in Canadian History and Environment (NiCHE), and others have served to connect scholars and to stimulate conversation and exchange. If done right, such networks can be a tremendous boost to countries and regions with many scattered researchers and few institutions with a critical mass of scholars working on related topics. NiCHE is the best current example of how a network can generate momentum, but I also believe that we have achieved some success in the Nordic countries through the Nordforsk network. Our future challenge is to manage to channel this increased exchange and communication into better, richer, deeper, and more exciting scholarship. Edited volumes such as this one can potentially serve not just as a physical manifestation of the network but also as a meeting place and a trading zone for new ideas and perspectives. To achieve this, however, such projects need to be based on sustained conversation and cooperation, and require strong editors with a clear idea of the project.

What Is the North? Where Is the North?

The idea of the North holds a special place in our understanding of the world, as the chapters in this volume make evident. But they also show that perhaps it is better to speak of the many ideas of many Norths. The North is a multitude of individual places; a geographical region comprising all these places; and also a plethora of cultural ideas, all arranged along a timeline. Like the magnetic north pole, the idea of North shifts around, evading simple categorization. The North evokes a series of binaries: north-south, centre-periphery, light-darkness, cold-warm, nature-culture, and so on, yet we should be aware of such simple binaries. Defining the North in terms of its real or imagined opposites falls short of providing a satisfactory definition. The authors demonstrate that what North means is by no means self-evident and absolute; it changes over time and between contexts. Furthermore, we see that different interpretations and understandings coexist simultaneously, in layers, with some in conflict, some in harmony, and others simply not in contact at all. Analyzing this interpretative flexibility and narratological ambiguity is an inspiration, I believe, for how we should think about the complex relationship between man and nature. For instance, the exploration of the North by the early heroic explorers was more than simply filling in the blank spaces on the map – it was also the construction of a particular kind of narrative about the relationship between man and nature, as Marionne Cronin's chapter in this volume clearly demonstrates. On another level, these activities and the narratives about them also contributed to the development of national identities and states. As the chapters by Seija A. Niemi and Ryan Tucker Jones reveal, the data gathered in the explorers' expeditions did in some cases become significant in the development of the sciences. And there are countless other narratives that could be told, of the meetings between the explorers and the people who already lived in the North, of the families that stayed behind, and so on.

Thinking about the past of the North as a region, an idea, and a multiplicity of lived experiences allows us to provide deeper, denser, and more meaningful analyses not only of the present but also of the future. Environmental history as a field has spent considerable effort in unravelling narratives of our environmental past, as well as bringing new narratives to the fore. Historical scholarship is, admittedly, a creative process, and not just reflexive. This volume is part of this re-evaluation of the North, particularly by extending the focus from equating the North with the Arctic, as so much scholarship has done, into thinking about the North in a broader sense.[3]

But if the North is not the same as the Arctic, then what is it? Where does the North begin? Can we pinpoint a particular place? Or time? Does such a question even make sense? Although there is no absolute answer to this question, the chapters in the volume all deal with places above a certain latitude, where particular environmental factors come into play. Weather, climate, and seasonality form a framework for life in the North. Winters tend to be dark and rich in snow; summers are short and bright. The population density is low, and certain geographic factors have also influenced the shape of communication and transport networks between settlements. But what seems like empty, pristine spaces are – if you extend them through history – in fact teeming with human and non-human activity and landscape modifications. How does this influence life and culture? How do people enable living in the North? All the chapters in the volume examine these questions in one way or another, though I think perhaps Simo Laakkonen's chapter strongly evokes the lived experience of the everyday in the North. The freezing-cold washwomen of Helsinki not only vividly illustrate the environmental conditions of life in the North but also remind us that these conditions are often unequally distributed.

Can the North be identified as a particular region, with specific environmental characteristics? Several chapters address this question, directly or indirectly. One way of thinking about this is to look at what environmental historians write about. This collection has no contributions from or about Denmark proper, which suggests that some regional concepts like the Nordic countries or Scandinavia have more to do with shared history and culture than environmental factors. Of course, Denmark has a long colonial history involving Greenland, the Faroe Islands, Iceland, and even Norway. These other concepts are tied to nation-states, and I would say that the North is something that transcends this level of organization – which is a key idea for the volume, beautifully captured in the title *Northscapes*. The North certainly plays a part in the complex construction of national identities, as Jones demonstrates, though I think this is a relationship that needs further exploration within environmental history. We are increasingly exploring the cultural and material processes in which species and ecosystems come to "belong" in particular places, becoming part of both the national culture and the national nature, but we have written less about the relationship between national states, identities, and natures.[4] This is beginning to change internationally, as seen in David Biggs's price-winning book *Quagmire*, which examines nation building and nature in the Mekong Delta.[5] At the same time, it is important to avoid environmental determinism, as Bathsheba Demuth's chapter shows us. Similar environmental conditions do not necessarily create similar developments, nor do particular landscapes automatically lead to particular national and cultural traits.

The creation, maintenance, and enforcement of regional boundaries are in many cases intensely related to environmental concerns and resources. Somewhat surprisingly, none of the chapters in this book deals with military activity and technology in the North. Despite some recent international research, the relationship between environmental history and military history is still underexplored.[6] Nor do they sufficiently capture the diplomatic history of the relationship between the various countries and regions in the North and the rest of the world. Environmental historians know that the natural boundaries between particular regions are more porous than political boundaries seem to be, as evidenced by the acid rain discussion of the 1980s, where Norwegian and Swedish forests turned out to be particularly vulnerable to pollution from British factories.[7]

The North implies a distance from the centre – in the Scandinavian case, from mainland Europe. This distance matters – something that, for instance, was continually reiterated during the Norwegian EU discussions of 1972 and 1994. Phrases like "the distance between Oslo and Tromsø is greater than the distance between Oslo and Brussels" served to remind people that even the European periphery of Norway has its own periphery. In the Introduction, Dolly Jørgensen and Sverker Sörlin set up the North as an imagined space, as imagined both by people who live there and people who don't, a topic that both Lisa Cooke and Unnur Birna Karlsdóttir explore in depth in their chapters. The communally imagined character of the North as a special place with special needs is critical to understanding the relationship between periphery and centre.

Seen from the centre, the North is certainly placed on the periphery of the world. The North becomes part of the vast hinterland that delivers goods and resources to the centre, as William Cronon so effectively describes in *Nature's Metropolis*.[8] From the North flow natural resources like fish, oil, iron, and timber to the rest of the world. Yet, we need to remember that the North has its own hinterland, drawing in resources and commodities from afar. While delving into historical travel accounts from obscure and peripheral places in the Scandinavian countryside from the early 1800s, one of the most fascinating things I noticed is how widespread coffee consumption was. Even in the most remote Norwegian mountain villages, poor farmers would roast and drink their own coffee on a daily basis. This astounded many British travellers who came here seeking what W. Cecil Slingsby termed "the Northern Playground," and demonstrates how even the northern periphery stands at the centre of its own hinterland, drawing on resources from all over the world.[9] Jan Kunnas's exploration of traversal technology transfer in the periphery provides another corrective to the centre-periphery model.

We know that, historically, the North seemed like a perfect target for "modernization."[10] It is a region of extensive natural riches, with waters teeming with life and forests full of trees. The North has long had such a dual identity, being both an area with extensive nature areas of critical importance for ecosystems and a site of immense natural resources. Some areas are sites for tourism and leisure lifestyles, supporting outdoor activities such as leisure cabin stays, camping, hunting, and skiing. At the same time, above- and underground industrial resources are plentiful; trees, grazing land, water, wind, and minerals attract numerous production industries to the North. In all of these cases, nature is consumed in one form or another. No wonder that, historically, much optimism has been connected to the North, both as a source of natural resources for industrial development and as a site of nature-based leisure. Yet, recent intensive industrial and extractive development such as oil and gas prospecting in northern Norway, mining in northern Sweden, and tar sands extraction and fracking in Canada has put much of this idea of pristine nature under pressure, causing many dystopian scenarios for northern nature. This tension between northern nature as a resource in itself – a reserve set aside from the modern world – and northern nature as a storehouse of resources that we can use to run the modern world, will remain key to understanding the idea of North.

The British tourists were among the first to seek out the Scandinavian countryside, and many published extensive travel accounts and guides about their travels. They were later joined by Norwegian and Swedish urbanites wanting to experience a more pure and authentic country. Particularly in Norway, landscape, culture, and national identity bled into each other to create a new idea of what it meant to be Norwegian following the country's independence from Denmark in 1814. At the same time, places like northern Norway and Sweden were sparsely populated, and the people were mostly poor, uneducated, and easily categorized as "backward" by the more "civilized" travellers from Britain or more urban national areas.[11] We can find similar stories and cultural stereotypes about Northerners from the entire circumpolar region. Moving into the twentieth century, the North continued to feature in many people's imaginations as a pure and more authentic space, far away from the polluted and densely populated regions of mainland Europe. To be sure, there are some significant Orientalism processes at work in the global imagination of the North, to borrow a concept from Edward Said.[12] Some went as far as to propose Norway as the national park of Europe, where Norway could remain so-called untouched wilderness while the rest of Europe continued urbanizing and industrializing.[13] But, of course, even before Norway discovered oil and became an affluent country, Norwegian nature was not untouched at all. It was a made environment, modified and experienced through technology and other human practices, a domesticated space connected, as I mentioned, through trade, travel, and information networks to the rest of the world.[14]

This continual mythologization of the North could perhaps serve as an analogue to Frederick Jackson Turner's frontier thesis.[15] For the young American nation, the constant move westward became the defining character of Americans, what distinguished them from the Old World they originally came from. In Europe, it seems like the North has played a similar role in the cultural politics of defining who we are and who we are not. However, as the chapter by Anna Gudrún Thórhallsdóttir, Árni Daníel Júlíusson, and Helga Ögmundardóttir indicates, the frontier is a complex idea that does not translate completely between different national contexts; it can refer to commodity extraction, to settlement patterns, and to an imagined nature. In a fully circumpolar perspective, the northern frontier becomes even more multi-faceted and contradictory.

Looking Toward the Future

To come back to the North, I think that we definitely haven't written the last word about the environmental histories of the North. There seems to be a sense of urgency in the way we speak of the North today, as global climate change and the melting polar ice loom large in the public imagination. The imaginary North is not supposed to be just a place outside history, as Dolly Jørgensen and Sverker Sörlin mention in their Introduction; it should also be a timeless and unchanging place. But we can no longer deny that change is coming to the North. In coming to terms with this change, we are forced to articulate what North is. I believe that the North will continue to face the challenges of balancing all these activities and historical layers of meaning, which makes knowledge mobilization ever more necessary. The northern regions of the world are poised for significant economic expansion in the twenty-first century, with increasing demands for biofuels, base and rare metals, and tourist facilities. Although tourism becomes more and more commercially important, the extractive industries are also growing.

There is a narrative arc in the West of moving away from polluting industries toward "softer" knowledge and experience economies (while shifting actual production and extraction over to other parts of the world), but this is not the case with mining. Industry is stronger than ever and currently expanding in northern Scandinavia because of the growing international demand for both base and exotic metals. Many of these materials are heavily used in the computing industry, but demand has also skyrocketed because of their use in electric automobile batteries and wind turbines, creating interesting connections between mining, consumption, and environmentally friendly practices. The mining industry has generated enormous environmental problems, transforming many areas of the rural North into industrialized landscapes.[16] The many layers of history and human activity become particularly visible in these areas, where the industrial and the post-industrial meet, as hybrid landscapes, comprising nature and culture, technology, and environment, future and past.[17]

At the same time, the many-layered complexity will also ensure that the North will remain a dynamic place central in global discussions and development. Environmental history can and should contribute to these discussions by highlighting three key insights from our scholarship:

- **The North is a *networked region*:** Northern places can't be understood as disconnected sites, isolated from the world; they are instead nodes, tightly networked and connected in a variety of ways. By using methods from geography, environmental history, history of technology, science and technology studies, and other relevant disciplines, scholars should continue to explore the flow of people, resources, and cultural meaning between northern sites and the rest of the world, and the technological infrastructures and mobility practices that make these connections possible.
- **The North is a *hybrid landscape*:** An extraordinary amount of scholarship has explored the idea of wilderness as a resource and how it needs to be protected from its opposite, a built environment that has been physically changed by human use. Environmental historians should avoid this dichotomy by thinking of the North as *hybrid landscapes*, neither urban nor wilderness, conquered nor preserved, but somewhere in between, as Richard White urges us to.[18] The North is a region shared between many different practices that overlap in space and over time. The coexistence of these practices forms historical layers – cultural, political, and material – that shape northern practices and communities. In other words, we can't discuss the future of the North without understanding its present,

and how the obduracy of previous practices and infrastructures create certain trajectories.

- **The North is a *site of consumption*:** The North is an integral part of modern consumer society. As part of global commodity chains, resources extracted from rural areas become consumer goods and industrial products that later return to the rural North. Over the course of the late nineteenth and early twentieth century, this development was particularly marked, as part of the transition to a currency-based local economy in the countryside. At the same time, the rural North has become a site of leisure consumption, especially through tourism. Environmental historians would do well to examine in more depth how these consumptive practices create connections between urban and rural, local and global.

In short, we need to move beyond the idea of the North as simply, or originally, pristine nature with wide spaces with few people and full of moose and mosquitos, while we simultaneously consider why this image of the North persists. There's nothing special to the North in that it is not something apart from the world we live in. Having lived in the North more or less all my life, it becomes everyday and ordinary. This is a necessary corrective to the mythologizing narratives of the North. Yet, at the same time, there's a value to the outside view, to making the North strange and new again. I hope this volume has succeeded in doing both.

NOTES

1 These larger meetings are complemented by a number of smaller and more focused international workshops, such as the "Bringing STS into Environmental History" meeting in Trondheim, Norway, in August 2010, the results of which were published as Dolly Jørgensen, Finn Arne Jørgensen, and Sara B. Pritchard, eds., *New Natures: Joining Environmental History with Science and Technology Studies* (Pittsburgh: University of Pittsburgh Press, 2013).

2 One example of such a re-examination was recently undertaken by a group of environmental historians representing five Nordic countries: Finn Arne Jørgensen, Unnur Birna Karlsdóttir, Erland Mårald, Bo Poulsen, and Tuomas Räsänen, "Entangled Environments: Historians and Nature in the Nordic Countries," *Historisk Tidsskrift* (Norway) 92 (2013): 9-34.

3 Bravo and Sörlin, *Narrating the Arctic*.

4 One exception is the interdisciplinary Nature and Nation network, which has organized several workshops and conference sessions to explore the relationship between nation-states and the environment. See http://www.natureandnation.eu/ for more information.

5 David Biggs, *Quagmire: Nation-Building and Nature in the Mekong Delta* (Seattle: University of Washington Press, 2010).

6 Edmund Russell, *War and Nature: Fighting Humans and Insects with Chemicals from World War I to Silent Spring* (Cambridge: Cambridge University Press, 2001); Richard P. Tucker and Edmund Russell, eds., *Natural Enemy, Natural Ally: Toward an Environmental History of Warfare* (Corvallis: Oregon State University Press, 2004).

7 Lars J. Lundgren, *Acid Rain on the Agenda: A Picture of a Chain of Events in Sweden, 1966-1968* (Lund, Sweden: Lund University Press, 1998).

8 William Cronon, *Nature's Metropolis: Chicago and the Great West* (New York: W.W. Norton, 1991).

9 William Cecil Slingsby, *Norway, the Northern Playground: Sketches of Climbing and Mountain Exploration in Norway Between 1872 and 1903* (printed by David Douglas, 10 Castle Street, Edinburgh, 1904).

10 Sverker Sörlin, *Framtidslandet: Debatten om Norrland och naturresurserna under det industriella genombrottet* (Stockholm: Carlssons Förlag, 1988).

11 One example is Henry David Inglis, *A Personal Narrative of a Journey through Norway, Part of Sweden, and the Islands and States of Denmark* (printed for Constable and Co., 1829). See also Peter Stadius, *Resan till norr: Spanska Nordenbilder kring sekelskiftet 1900*, Bidrag till kännedom av Finlands natur och folk [Contributions to the understanding of the nature and people of Finland] 164 (Helsinki: Finska Vetenskaps-Societeten, 2005), which demonstrates that southern European travellers also were impressed by some of the technological infrastructures they encountered in the Nordic countries, which clashed with their preconceptions of the primitive North.

12 Edward Said, *Orientalism* (New York: Vintage Books, 1979).

13 Erik Langdalen and Karl Erikstad, *Plan i fjellet/Hytta på fjellet* (Oslo: Grøndahl & Søn, 1966), 1.

14 I have myself explored how leisure activities have transformed Norwegian nature into a built environment. See Finn Arne Jørgensen, "Den første hyttekrisa: Samfunnsplanlegging, naturbilder og allmenningens tragedie," in *Norske hytter i endring: Om bærekraft og behag*, ed. Helen Jøsok Gansmo, Thomas Berker, and Finn Arne Jørgensen (Trondheim: Tapir Academic Press, 2011), 37-52.

15 Frederick Jackson Turner, *The Frontier in American History* (Mineola, NY: Dover, 1996).

16 See, for example, Liza Piper, *The Industrial Transformation of Subarctic Canada* (Vancouver: UBC Press, 2010), and Timothy LeCain, *Mass Destruction: The Men and Giant Mines That Wired America and Scarred the Planet* (Piscataway, NJ: Rutgers University Press, 2009).

17 See, for example, Jan Jörnmark, *Övergivna platser* (Lund, Sweden: Historiske Media, 2007) or Edward Burtynsky, *Manufactured Landscapes: The Photographs of Edward Burtynsky* (New Haven, CT: Yale University Press, 2003) for powerful photographic explorations of such landscapes.

18 Richard White, "From Wilderness to Hybrid Landscapes: The Cultural Turn in Environmental History," *Historian* 66 (2004): 557-64.

Selected Bibliography

Anderson, David, and Mark Nutall, eds. *Cultivating Arctic Landscapes: Knowing and Managing Animals in the Circumpolar North*. New York: Berghahn Books, 2004.

Arnold, David. *The Problem of Nature: Environment, Culture and European Expansion*. Oxford: Blackwell, 1995.

Baron, Nick. *Soviet Karelia: Politics, Planning and Terror in Stalin's Russia, 1920-1939*. London: Routledge, 2007.

Barrett, James H., ed. *Contact, Continuity and Collapse: The Norse Colonization of the North Atlantic*. Turnhout, Belgium: Brepols, 2003.

Basalla, George. *The Evolution of Technology*. Cambridge: Cambridge University Press, 1988.

Batey, Colleen, Judith Jesch, and Christopher D. Morris, eds. *The Viking Age in Caithness, Orkney and the North Atlantic*. Edinburgh: Edinburgh University Press, 1993.

Benson, Keith R., and Helen M. Rozwadowski, eds. *Extremes: Oceanography's Adventures at the Poles*. Sagamore Beach, MA: Science History Publications USA, 2007.

Bertelsen, Reidar. "Farm Mounds in North Norway: A Review of Recent Research." *Norwegian Archaeological Review* 12 (1979): 48-56.

Berton, Pierre. *The Impossible Railway: The Building of the Canadian Pacific*. New York: Alfred A. Knopf, 1972.

Bhabha, Homi. *Nation and Narration*. New York: Routledge, 1990.

Biggs, David. *Quagmire: Nation-Building and Nature in the Mekong Delta*. Seattle: University of Washington Press, 2010.

Bloom, Lisa. *Gender on Ice: American Ideologies of Polar Expeditions.* Minneapolis: University of Minnesota Press, 1993.

Bocking, Stephen. "A Disciplined Geography: Aviation, Science, and the Cold War in Northern Canada, 1945-1960." *Technology and Culture* 50, 2 (2009): 265-90.

–. "Science and Spaces in the Northern Environment." *Environmental History* 12, 4 (2007): 867-94.

Bolotova, Alla. "Colonization of Nature in the Soviet Union: State Ideology, Public Discourse, and the Experience of Geologists." *Historical Social Research* 29, 3 (2004): 104-23.

Bowker, Geoffrey C., and Susan Leigh Star. *Sorting Things Out: Classification and Its Consequences.* Cambridge, MA: MIT Press, 1999.

Braun, Bruce. "Place Becoming Otherwise." *BC Studies* 131 (2001): 15-24.

–. *The Intemperate Rainforest: Nature, Culture, and Power on Canada's West Coast.* Minneapolis: University of Minnesota Press, 2002.

Bravo, Michael T., and Sverker Sörlin. *Narrating the Arctic: A Cultural History of Nordic Scientific Practices.* Canton, MA: Science History Publications, 2002.

Breidenbach, Joanna, and Pál Nyíri. "'Our Common Heritage': New Tourist Nations, Post-'Socialist' Pedagogy, and the Globalization of Nature." *Current Anthropology* 48 (2007): 322-30.

Breyfogle, Nicholas B., Abby Schrader, and Willard Sunderland, eds. *Peopling the Russian Periphery: Borderland Colonization in Eurasian History.* London: Routledge, 2007.

Browne, Janet. *The Secular Ark: Studies in the History of Biogeography.* New Haven, CT, and London: Yale University Press, 1983.

Bruno, Andy. "Industrial Life in a Limiting Landscape: An Environmental Interpretation of Stalinist Social Conditions in the Far North." *International Review of Social History* 55 (2010): 153-74.

Carr, Edward Hallett. *What Is History? The George Macaulay Trevelyan Lectures Delivered at the University of Cambridge, January-March 1961.* New York: Vintage Books, 1961.

Coates, Kenneth, and William R. Morrison. *Land of the Midnight Sun: A History of the Yukon.* Montreal and Kingston: McGill-Queen's University Press, 2005.

Corn, Joseph J. *The Winged Gospel: America's Romance with Aviation, 1900-1950.* New York: Oxford University Press, 1983.

Cosgrove, Denis, and Veronica della Dora, eds. *High Places: Cultural Geographies of Mountains, Ice and Science.* London: I.B. Tauris, 2009.

Cronon, William. *Nature's Metropolis: Chicago and the Great West.* New York: W.W. Norton, 1991.

–. "The Trouble with Wilderness; or, Getting back to the Wrong Nature." In *Uncommon Ground: Toward Reinventing Nature,* edited by William Cronon, 69-90. New York: W.W. Norton, 1995.

Daniels, Stephen. *Fields of Vision: Landscape Imagery and National Identity in England and the United States.* Princeton, NJ: Princeton University Press, 1993.

Diamond, Jared. *Collapse: How Societies Choose to Fail or Succeed.* New York: Penguin, 2005.

Driver, Felix. *Geography Militant: Cultures of Exploration and Empire.* Oxford: Blackwell, 2001.

Edgerton, David. *The Shock of the Old: Technology and Global History since 1900.* New York and Oxford: Oxford University Press, 2006.

Ely, Christopher. *This Meager Nature: Landscape and National Identity in Imperial Russia.* DeKalb: Northern Illinois University Press, 2002.

Emmerson, Charles. *The Future History of the Arctic: How Climate, Resources and Geopolitics Are Reshaping the North, and Why It Matters to the World.* London: Vintage Books, 2010.

Farish, Matthew. "Creating Cold War Climates: The Laboratories of American Globalism." In *Environmental Histories of the Cold War,* edited by John R. McNeill and Corinna R. Unger, 51-84. Cambridge: Cambridge University Press, 2010.

Fèbvre, Lucien. *La Terre et l'évolution humaine.* Paris: Albin Michel, 1922.

Flanagan, Maureen. *Seeing with Their Hearts: Chicago Women and the Vision of a Good City, 1877-1933.* Princeton, NJ: Princeton University Press, 2002.

Folke Ax, Christina, Niels Brimnes, Niklas Thode Jensen, and Karen Oslund, eds. *Cultivating the Colony: Colonial States and Their Environmental Legacies.* Athens: Ohio University Press, 2011.

Friedel, Robert A. *Culture of Improvement: Technology and the Western Millennium.* Cambridge, MA: MIT Press, 2010.

Frost, Alan. "The Pacific Ocean: The Eighteenth Century's 'New World.'" *Studies on Voltaire and the Eighteenth Century* 152 (1976): 779–822.

Fur, Gunlög. *Colonialism in the Margins – Cultural Encounters in New Sweden and Lapland.* Leiden and Boston: Brill, 2006.

Gansum, Terje, and Terje Oestigaard. "The Ritual Stratigraphy of Monuments That Matter." *European Journal of Archaeology* 7 (2004): 61-79.

Giddens, Anthony. *The Constitution of Society: Outline of the Theory of Structuration.* Cambridge, UK: Polity Press, 1984.

Glacken, Clarence. *Traces on the Rhodian Shore.* Berkeley: University of California Press, 1967.

Gordon, Avery F. *Ghostly Matters: Haunting and the Sociological Imagination.* 2nd ed. Minneapolis and London: University of Minnesota Press, 2008.

Grace, Sherrill E. *Canada and the Idea of the North.* Montreal and Kingston: McGill-Queen's University Press, 2001.

Griffiths, Tom, and Libby Robin, eds. *Ecology & Empire: Environmental History of Settler Societies.* Edinburgh: Keele University Press, 1997.

Grove, Richard. "Environmental History." In *New Perspectives on Historical Writing.* 2nd ed., edited by Peter Burke, 261-82. University Park: Pennsylvania State University Press, 2004.

Gugliotta, Angela. "Class, Gender, and Coal Smoke: Gender Ideology and Environmental Injustice in Pittsburgh, 1868-1914." *Environmental History* 5 (2000): 165-93.

Gupta, Akhil, and James Ferguson, eds. *Culture, Power, Place: Explorations in Critical Anthropology*. Durham, NC, and London: Duke University Press, 1997.

Hálfdanarson, Guðmundur. "The Nordic Area: From Competition to Cooperation." In *Empires and States in European Perspective*, edited by Steven Ellis, 83-93. Pisa: Edizioni Plus, 2002.

Hallsdóttir, Margrét. "On Pre-Settlement History of Icelandic Vegetation." *Búvísindi* 9 (1995): 17-29.

Headrick, Daniel R. *The Tentacles of Progress: Technology Transfer in the Age of Imperialism, 1850-1940*. New York and Oxford: Oxford University Press, 1988.

Hevly, Bruce. "The Heroic Science of Glacier Motion." *Osiris* 11 (1996): 66-80.

Heymann, Matthias, Henrik Knudsen, Maiken L. Lolck, Henry Nielsen, Kristian H. Nielsen, and Christopher J. Ries. "Exploring Greenland: Science and Technology in Cold War Settings." *Scientia Canadensis* 33 (2010): 11-42.

Hill, J.D. *Ritual and Rubbish in the Iron Age of Wessex: A Study on the Formation of a Specific Archaeological Record*. Oxford: Tempus Reparatum, 2005.

Hill, Jen. *White Horizon: The Arctic in the Nineteenth-Century British Imagination*. Albany: State University of New York Press, 2008.

Holquist, Peter. "'In Accord with State's Interests and the People's Wishes': The Technocratic Ideology of Imperial Russia's Resettlement Administration." *Slavic Review* 69, 1 (2010), 152.

Horensma, Pier. *The Soviet Arctic*. London: Routledge, 1991.

Hughes, J. Donald. *What Is Environmental History?* Cambridge, UK: Polity Press, 2006.

Hughes, Thomas P. *Human-Built World: How to Think about Technology and Culture*. Chicago: University of Chicago Press, 2004.

Ingold, Tim. *The Appropriation of Nature: Essays on Human Ecology and Social Relations*. Manchester: Manchester University Press, 1986.

–. *The Perception of the Environment: Essays on Livelihood, Dwelling and Skill*. London: Routledge, 2000.

Jakobsson, Sverrir, ed. *Images of the North: Histories, Identities, Ideas*. Amsterdam and New York: Rodopi, 2009.

Jones, Ryan. "'A Havock Made among Them': Animals, Empire, and Extinction in the Russian North Pacific, 1741-1810." *Environmental History* 16 (2011): 585-609.

–. "Lisiansky's Mountain: Changing Views of Russian America." *Alaska History* 25 (2010): 1-22.

Jørgensen, Dolly, Finn Arne Jørgensen, and Sara Pritchard, eds. *New Natures: Joining Environmental History with Science and Technology Studies*. Pittsburgh: University of Pittsburgh Press, 2013.

Jørgensen, Finn Arne. *Making a Green Machine: The Infrastructure of Beverage Container Recycling*. New Brunswick, NJ: Rutgers University Press, 2011.

Jørgensen, Finn Arne, Unnur Birna Karlsdóttir, Erland Mårald, Bo Poulsen, and Tuomas Räsänen. "Entangled Environments: Historians and Nature in the Nordic Countries." *Historisk Tidsskrift* (Norway) 92 (2013): 9-34.

Kangaspuro, Markku, and Jeremy Smith, eds. *Modernization in Russia since 1900*. Helsinki: Finnish Literature Society, 2006.

Keskitalo, E.C.H. *Negotiating the Arctic: The Construction of an International Region*. New York: Routledge, 2004.

Kimmel, Michael S. *Manhood in America: A Cultural History.* 2nd ed. Oxford: Oxford University Press, 2006.

Kish, George. "Adolf Erik Nordenskiöld (1832-1901): Polar Explorer and Historian of Cartography." *Geographical Journal* 134, 4 (1968): 487-500.

Krupnik, Igor. *Arctic Adaptations: Native Whalers and Reindeer Herders of Northern Eurasia.* Translated by Marcia Levenson. Hanover, NH: Dartmouth College Press, 1993.

Kunnas, Jan. "A Dense and Sickly Mist from Thousands of Bog Fires." *Environment and History* 11, 4 (2005): 431-46.

Lajus, Julia A. "'Foreign Science' in a Russian Context: Murman Scientific-Fishery Expedition and Russian Participation in Early ICES Activity." *ICES Marine Science Symposia* 215 (2002): 64-72.

Launius, Roger, James Rodger Fleming, and Donald H. Devorkin, eds. *Globalizing Polar Science: Reconsidering the International Polar and Geophysical Years.* New York: Palgrave, 2010.

Lawson, I.T., F.J. Gathorne-Hardy, M.J. Church, A.J. Newton, K.J. Edwards, A.J. Dugmore, and A. Einarsson. "Environmental Impacts of the Norse Settlement: Palaoenvironmental Data from Mývatnssveit, Northern Iceland." *Boreas* 36 (2007): 1-19.

Lears, T.J. Jackson. *No Place of Grace: Antimodernism and the Transformation of American Culture, 1880-1920.* New York: Pantheon Books, 1981.

LeCain, Timothy. *Mass Destruction: The Men and Giant Mines That Wired America and Scarred the Planet.* New Brunswick, NJ: Rutgers University Press, 2009.

Levere, Trevor. *Science and the Canadian Arctic: A Century of Exploration, 1818-1918.* Cambridge: Cambridge University Press, 1993.

Loo, Tina. "Of Moose and Men: Hunting for Masculinities in British Columbia, 1880-1939." *Western Historical Quarterly* 32, 3 (2001): 296-319.

Lundgren, Lars J. *Acid Rain on the Agenda: A Picture of a Chain of Events in Sweden, 1966-1968.* Lund, Sweden: Lund University Press, 1998.

Malcolmson, Patricia. *English Laundresses: A Social History, 1850-1930.* Urbana: University of Illinois Press, 1986.

McCannon, John. *A History of the Arctic: Nature, Exploration, and Exploitation.* London: Reaktion Books, 2013

–. *Red Arctic: Polar Exploration and the Myth of the North in the Soviet Union.* New York and Oxford: Oxford University Press, 1998.

Melosi, Martin, ed. *Pollution and Reform in American Cities, 1870-1930.* Austin: University of Texas Press, 1980.

Merchant, Carolyn, ed. *Women and Environmental History.* Special issue, *Environmental History Review*, Spring 1984: 1-112.

Mergen, Bernard. *Snow in America.* Washington, DC: Smithsonian Institution Press, 1979.

Misa, Thomas J. *Leonardo to the Internet: Technology and Culture from the Renaissance to the Present.* Baltimore: Johns Hopkins University Press, 2004.

Mohun, Arwen Pelmer. "Laundrymen Construct Their World: Gender and the Transformation of a Domestic Task to an Industrial Process." *Technology and Culture* 38 (1997): 97-120.

Moore, Jason W. "'Amsterdam Is Standing on Norway' Part I: The Alchemy of Capital, Empire and Nature in the Diaspora of Silver, 1545-1648." *Journal of Agrarian Change* 10 (2010): 33-68.

–. "'Amsterdam Is Standing on Norway' Part II: The Global North Atlantic in the Ecological Revolution of the Long Seventeenth Century." *Journal of Agrarian Change* 10 (2010): 188-277.

Morse, Kathryn. *The Nature of Gold: An Environmental History of the Klondike Gold Rush*. Seattle and London: University of Washington Press, 2003.

Mosley, Stephen. "Common Ground: Integrating Social and Environmental History." *Journal of Social History* 39, 3 (2006): 915-33.

Moss, Sarah. *The Frozen Ship: The Histories and Tales of Polar Exploration*. New York: Blue Bridge, 2006.

Myllyntaus, Timo, Minna Hares, and Jan Kunnas, "Sustainability in Danger? Slash-and-Burn Cultivation in Nineteenth-Century Finland and Twentieth-Century Southeast Asia." *Environmental History* 7, 2 (2002): 267-302.

Nash, Roderick. *Wilderness and the American Mind*. New Haven, CT: Yale University Press, 1967.

Needham, Stuart, and Tony Spence. "Refuse and the Formation of Middens." *Antiquity* 71 (1997): 77 90.

Nye, Robert A. "Medicine and Science as Masculine 'Fields of Honor.'" *Osiris* 12 (1997): 60-79.

O'Connor, Ralph. *Earth on Show: Fossils and the Poetics of Popular Science, 1802-1856*. Chicago: University of Chicago Press, 2008.

Osborne, Brian S. "Landscape, Memory, Monuments, and Commemoration: Putting Identity in Its Place." *Canadian Ethnic Studies* 33 (2001): 39-77.

Osherenko, Gail, and Oran R. Young, *The Age of the Arctic: Hot Conflicts and Cold Realities*. Cambridge: Cambridge University Press, 1989.

Oslund, Karen. *Iceland Imagined: Nature, Culture, and Storytelling in the North Atlantic*. Seattle: University of Washington Press, 2011.

Owen, Olwyn, ed. *The World of Orkneyinga Saga: "The Broad-Cloth Viking Trip."* Kirkwall, UK: Orcadian, 2005.

Piper, Liza. *The Industrial Transformation of Subarctic Canada*. Vancouver: UBC Press, 2009.

Pisano, Dominick A., ed. *The Airplane in American Culture*. Ann Arbor: University of Michigan Press, 2003.

Platt, Harold L. "Invisible Gases: Smoke, Gender, and the Redefinition of Environmental Policy in Chicago, 1900-1920." *Planning Perspectives* 10 (1995): 67-97.

Pretes, M. "Tourism and Nationalism," *Annals of Tourism Research* 30 (2003): 125-42.

Pritchard, Sara. *Confluence: The Nature of Technology and the Remaking of the Rhône*. Cambridge, MA: Harvard University Press, 2011.

Pyne, Stephen J. *The Ice: A Journey to Antarctica*. Iowa City: University of Iowa Press, 1986.

–. *Vestal Fire – An Environmental History, Told through Fire, of Europe and Europe's Encounter with the World.* Seattle: University of Washington Press, 1997.

Radkau, Joakim. *Nature and Power: A Global History of the Environment.* Cambridge: Cambridge University Press, 2008.

Reuss, Martin, and Stephen Cutcliffe, eds. *Illusory Boundary: Technology and the Environment.* Charlottesville: University of Virginia Press, 2010.

Riffenburgh, Beau. *The Myth of the Explorer: The Press, Sensationalism, and Geographical Discovery.* London: Belhaven Press, 1993.

Ritvo, Harriet. *The Platypus and the Mermaid: And Other Figments of the Classifying Imagination.* Cambridge, MA: Harvard University Press, 1998.

Robinson, Michael F. *The Coldest Crucible: Arctic Exploration and American Culture.* Chicago: University of Chicago Press, 2006.

Rudwick, Martin J.S. *The Meaning of Fossils: Episodes in the History of Palaeontology.* Chicago and London: University of Chicago Press, 1985.

Russell, Edmund. *War and Nature: Fighting Humans and Insects with Chemicals from World War I to Silent Spring.* Cambridge: Cambridge University Press, 2001.

Said, Edward. *Orientalism.* New York: Vintage Books, 1979.

Sandlos, John. *Hunters at the Margin: Native People and Wildlife Conservation in the Northwest Territories.* Vancouver: UBC Press, 2007.

Sarmela, Matti. "Swidden Cultivation in Finland as a Cultural System." *Suomen antropologi* 12, 4 (1987): 241-49.

Scharff, Virginia, ed. *Seeing Nature through Gender.* Lawrence: University Press of Kansas, 2003.

Schrepfer, Susan, and Philip Scranton, eds. *Industrializing Organisms: Introducing Evolutionary History.* New York: Routledge, 2003.

Scott, James C. *Seeing Like a State: How Certain Schemes to Improve the Human Condition Have Failed.* New Haven, CT: Yale University Press, 1999.

Shadian, Jessica, and Monica Tennberg, eds. *Legacies and Change in Polar Science: Historical, Legal and Political Aspects on the International Polar Year.* Farnham, UK: Ashgate, 2010.

Shields, Rob. *Places on the Margin.* London: Routledge, 1991.

Simpson, I.A., A.J. Dugmore, A. Thomson, and O. Vesteinsson. "Crossing the Thresholds: Human Ecology and Historical Patterns of Landscape Degradation." *Catena* 42 (2001): 175-92.

Slezkine, Yuri. *Arctic Mirrors: Russia and the Small Peoples of the North.* Ithaca, NY: Cornell University Press, 1994.

Sörlin, Sverker. "Narratives and Counter Narratives of Climate Change: North Atlantic Glaciology and Meteorology, ca 1930-1955." *Journal of Historical Geography* 35 (2009): 237-55.

–. "The Anxieties of a Science Diplomat: Field Co-production of Climate Knowledge and the Rise and Fall of Hans Ahlmann's 'Polar Warming.'" *Osiris* 26 (2011): 66-88.

–, ed. *Science, Geopolitics and Culture in the Polar Region – Norden beyond Borders.* Farnham, UK: Ashgate, 2013.

Sörlin, Sverker, and Paul Warde, eds. *Nature's End: History and the Environment.* London: Palgrave Macmillan, 2009.

Spufford, Francis. *I May Be Some Time: Ice and the English Imagination.* London: Faber and Faber, 1996.

Stewart, Charles T., and Yasumitsu Nihei. *Technology Transfer and Human Factors.* Lexington, MA: Lexington Books, 1987.

Taylor, C.J. *Negotiating the Past: The Making of Canada's National Historic Parks and Sites.* Montreal and Kingston: McGill-Queen's University Press, 1990.

Tucker, Richard P., and Edmund Russell, eds. *Natural Enemy, Natural Ally: Toward an Environmental History of Warfare.* Corvallis: Oregon State University Press, 2004.

Turner, Frederick Jackson. *The Frontier in American History.* Mineola, NY: Dover Publications, 1996. First published in 1920 by H. Holt.

Uvachan, V.N. *The Peoples of the North and Their Road to Socialism.* Moscow: Progress, 1975.

Vance, Jonathan F. *Building Canada: People and Projects That Shaped the Nation.* Toronto: Penguin, 2006.

Wallerstein, Immanuel. *The Modern World System: Capitalist Agriculture and the Origins of the European World-Economy in the Sixteenth Century.* New York: Academic Press, 1974.

Webb, Melody. *Yukon: The Last Frontier.* Vancouver: UBC Press, 1993.

White, Richard. "From Wilderness to Hybrid Landscapes: The Cultural Turn in Environmental History." *Historian* 66 (2004): 557-64.

–. *The Organic Machine: The Remaking of the Columbia River.* New York: Hill and Wang, 1996.

Whited, Tamara, Jens Engels, Richard Hoffmann, Hilde Ibsen, and Wybren Verstegen. *Northern Europe: An Environmental History.* Santa Barbara: ABC Clio, 2005.

Willis, Roxanne. *Alaska's Place in the West: From the Last Frontier to the Last Great Wilderness.* Lawrence: University Press of Kansas, 2010.

Wolf, Eric R. *Europe and the People without History.* Berkeley, Los Angeles, and London: University of California Press, 1982.

Wood, Alan. *Russian's Frozen Frontier: A History of Siberia and the Russian Far East, 1581-1991.* NY: Bloomsbury Academic, 2011.

Wool, David. "Charles Lyell – 'the Father of Geology' – as a Forerunner of Modern Ecology." *OIKOS* 94, 3 (2001): 385-91.

Worster, Donald. *Dust Bowl: The Southern Plains in the 1930s.* New York: Oxford University Press, 1979.

Wynn, Graeme. *Canada and Arctic North America: An Environmental History.* Santa Barbara: ABC Clio, 2007.

Young, Oran R. *To the Arctic: An Introduction to the Far Northern World.* London: Wiley, 1989.

Yurchenko, Alexei, and Jens Petter Nielsen, eds. *In the North My Nest Is Made: Studies in the History of the Murman Colonization, 1860-1940.* St. Petersburg: European University at St. Petersburg and University of Tromso, 2005.

Zeller, Suzanne. *Inventing Canada: Early Victorian Science and the Idea of Transcontinental Nation*. Montreal: McGill-Queen's University Press, 2009.

Zeller, Thomas, and Christoph Mauch, eds. *The World beyond the Windshield: Roads and Landscapes in the United States and Europe*. Athens: Ohio University Press, 2008.

Contributors

Lisa Cooke is a cultural anthropologist specializing in Indigenous studies. Her research interests revolve around examining indigenous–settler relations in Canada. She has found ethnographic examinations of tourism and the production of touristic spaces a great entry point to exploring contemporary colonial cultural forms. To date, most of her work has been conducted in Whitehorse and Dawson City, in Canada's Yukon Territory. Lisa earned a PhD from York University and now calls Kamloops, British Columbia, home. She is an assistant professor of Anthropology at Thompson Rivers University.

Marionne Cronin is a research fellow in the Northern Colonialism Programme at the University of Aberdeen. Trained as a historian of technology, she is interested in the relationship between technology and place in the circumpolar world, particularly the history of circumpolar aviation and the role of technology in the history of polar exploration. Her current research explores how the technologies we use to move through northern environments leave traces on our ideas of "North."

Bathsheba Demuth is a doctoral candidate in history at the University of California, Berkeley, where her research focuses on the interactions of ecology and ideology in the nineteenth-and twentieth-century development of western Alaska and northeastern Russia. Before beginning graduate work in environmental history, she lived in the North American Arctic for several years, and served as a Peace Corps volunteer in the former Soviet Union. She holds an MA in International Development from Brown University.

Jane Harrison is an archaeologist at the University of Oxford. Her research interests range from the Viking settlement of the North Atlantic and the early medieval period in Britain to the theory and methodology of urban archaeology. She is currently finishing a PhD at Oxford, where she also teaches, runs a community archaeology research project in East Oxford, and is the excavation director of a project examining the Viking landscape of the Orkney Islands.

Ryan Tucker Jones is an assistant professor of history at Idaho State University. Among his publications are *Empire of Extinction: Russians and the Strange Beasts of the Sea in the North Pacific, 1709-1867* and several articles on Russian and Pacific environmental history. He is currently working on a global history of the Russian whaling industry.

Dolly Jørgensen is an environmental historian who has researched a broad array of topics, including medieval forestry management, late medieval urban sanitation, the modern practice of converting offshore oil structures into artificial reefs, and environmentalism in science fiction. She was a practising environmental engineer before earning a PhD in history from the University of Virginia. She is currently in the Department of Ecology and Environmental Science at Umeå University, Sweden, where she is researching the comparative history of animal reintroduction in Norway and Sweden. She is a co-editor of the volume *New Natures: Joining Environmental History with Science and Technology Studies.*

Finn Arne Jørgensen is an associate senior lecturer in History of Technology and Environment at Umeå University, Sweden. He holds a PhD in Science and Technology Studies from the Norwegian University of Science and Technology. He is the author of *Making a Green Machine: The Infrastructure of Beverage Container Recycling* and co-editor of *Norske hytter i endring: Om bærekraft og behag* and *New Natures: Joining Environmental History with Science and Technology Studies.* His current research includes a historical study of Norwegian leisure cabin culture.

Árni Daníel Júlíusson is a historian of peasant societies, which in practice has led to an effort to introduce methods of social and environmental history into the study of Icelandic history before 1800. His main research areas are pre-capitalist agricultural systems, ecological carrying capacity, and manorialism and pre-capitalist social conflict. He is one of three editors of the *Icelandic Historical Atlas*, originally issued from 1989 to 1993 and reissued and updated in 2005, and has written two volumes of the *Agricultural History of Iceland*. He holds a PhD from the University of Copenhagen and is an independent scholar at the ReykjavíkurAkademían.

Unnur Birna Karlsdóttir is an Icelandic historian whose primary focus is environmental history, but her research interests include the history of ideas and gender history, such as the history of abortion, eugenics, and racism in Iceland in a comparative light. She was a freelance researcher and an archivist at the National Archives of Iceland before earning a PhD in history from the University of Iceland. She is currently the curator of the East Iceland Hereditary Museum. She has published several articles and books in Icelandic.

Jan Kunnas earned his PhD in History and Civilization from the European University Institute in Florence, Italy. He has done extensive research on Finland's transition from a solar-based energy system to a fossil-fuel based one, and the environmental consequences of this transition. His is currently working at the University of Stirling, in Scotland, on a project that combines insights from economics and history to conduct long-run tests of the predictive power of indicators of sustainable development.

Simo Laakkonen is a Finnish environmental historian who earned his PhD in economic and social history from the University of Helsinki. His research has focused on urban environmental history, historical ecology of war, and environmental history of the Baltic Sea. He has directed several interdisciplinary research networks, which have included environmental engineers, natural scientists, social scientists, and historians who have explored these topics. His main publications include two special issues on the Baltic Sea published by the Royal Swedish Academy of Sciences. He is currently a university lecturer of Landscape Studies at the University of Turku, Finland.

Julia Lajus is an associate professor at the National Research University Higher School of Economics at St. Petersburg and at the European University at St. Petersburg, where she also leads the Centre for Environmental and Technological History. Her research interests include the history of field sciences, such as marine biology, oceanography, and geophysics, and the environmental history of biological resources in the Russian north in comparative and transnational perspectives. She holds a PhD in history from the Institute for the History of Science and Technology, Russian Academy of Sciences. She has published widely on the history of Soviet and Russian Arctic science and environment.

Seija A. Niemi is an environmental historian and non-fiction writer. She completed her MA in Cultural History at the University of Turku, Finland, with a thesis on the Finnish Tango King Olavi Virta, and earned her PhL in Finnish history with an environmental thesis on the history of the birch tree in Finnish forests. She recently finished her doctoral thesis on the Finnish-Swedish explorer and scientist Adolf Erik Nordenskiöld and his place in the history of Finnish conservation. She is also working on a series of non-fiction books for children about the Arctic.

Helga Ögmundardóttir is an anthropologist who has done research on various topics, including resource management conflicts in Iceland; socio-cultural characteristics of Icelandic fishing communities; interaction between Aboriginals and Icelandic immigrants in North America; and fuel transitions, energy management, and climate change adaptation in the Nordic countries. She has a PhD in cultural anthropology from Uppsala University, Sweden, and taught at the University of Iceland for nine years. She is currently a post-doctoral researcher at the Norwegian University of Science and Technology, with the Nordic research project NordStar.

Sverker Sörlin is a professor of Environmental History in the Division of History of Science, Technology and Environment at the KTH Royal Institute of Technology in Stockholm, where he also works with the KTH Environmental Humanities Laboratory. His current research is on the history of science politics of climate change and on the emergence and growth of environmental expertise in the twentieth century. Recent books include the co-edited *The Future of Nature: Documents of Global Change* and *Nature's End: History and the Environment*. Sörlin is a member of the Swedish government's Environmental Research Advisory Board.

Anna Gudrún Thórhallsdóttir is an Icelandic range ecologist with an MSc in natural resources from Norway and an MSc and PhD in range ecology from Utah State University. She is a professor at the Agricultural University of Iceland, where she also coordinates Natural History and Management Studies. Since 2010 she has worked for the Norwegian Institute for Agricultural and Environmental Research. Her research includes animal and plant aspects of grazing, landscape level effects, and the historical aspects of grazing. In 2009, she initiated a new interdisciplinary stream, Nature and History, at the Agricultural University of Iceland in cooperation with the University of Iceland.

Index

Note: "f" after a number indicates a figure; "m" after a number indicates a map

Printed and bound in Canada by Friesens

Set in Futura Condensed and Warnock
by Artegraphica Design Co. Ltd.

Copy editor: Judy Phillips

Proofreader: Kirsten Craven

Indexer: Patricia Buchanan

Cartographer: Eric Leinberger